D0025425

Table of Contents	Page

Table of Contents	Page

Table of Contents	Page

Table of Contents	Page

Project No. ____________

TITLE ____ Book No. ____________

From Page No. ____

To Page No. ____

Witnessed & Understood by me,	Date	Invented by:	Date
		Recorded by:	

Project No. ____________

Book No. ____________ TITLE ____________

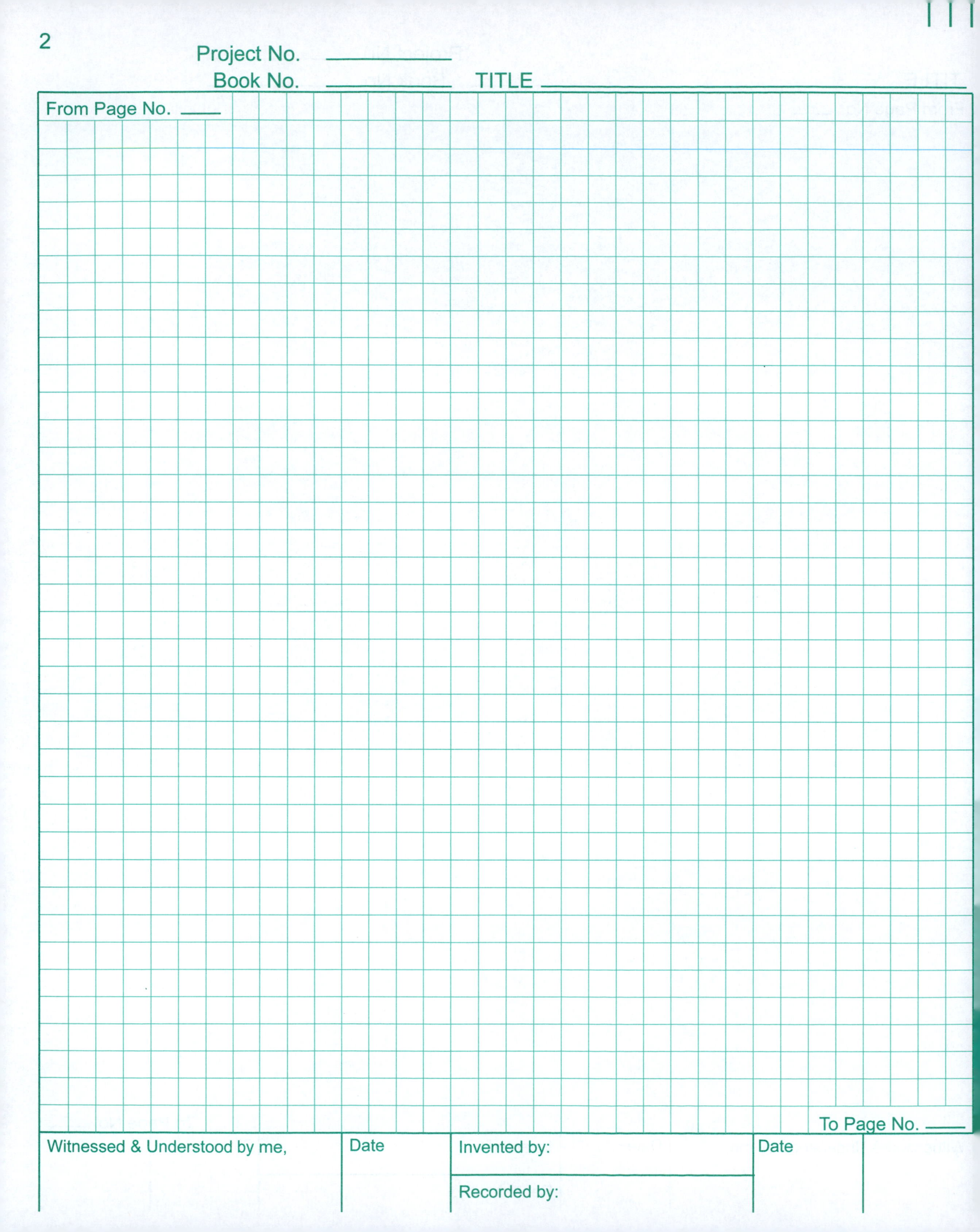

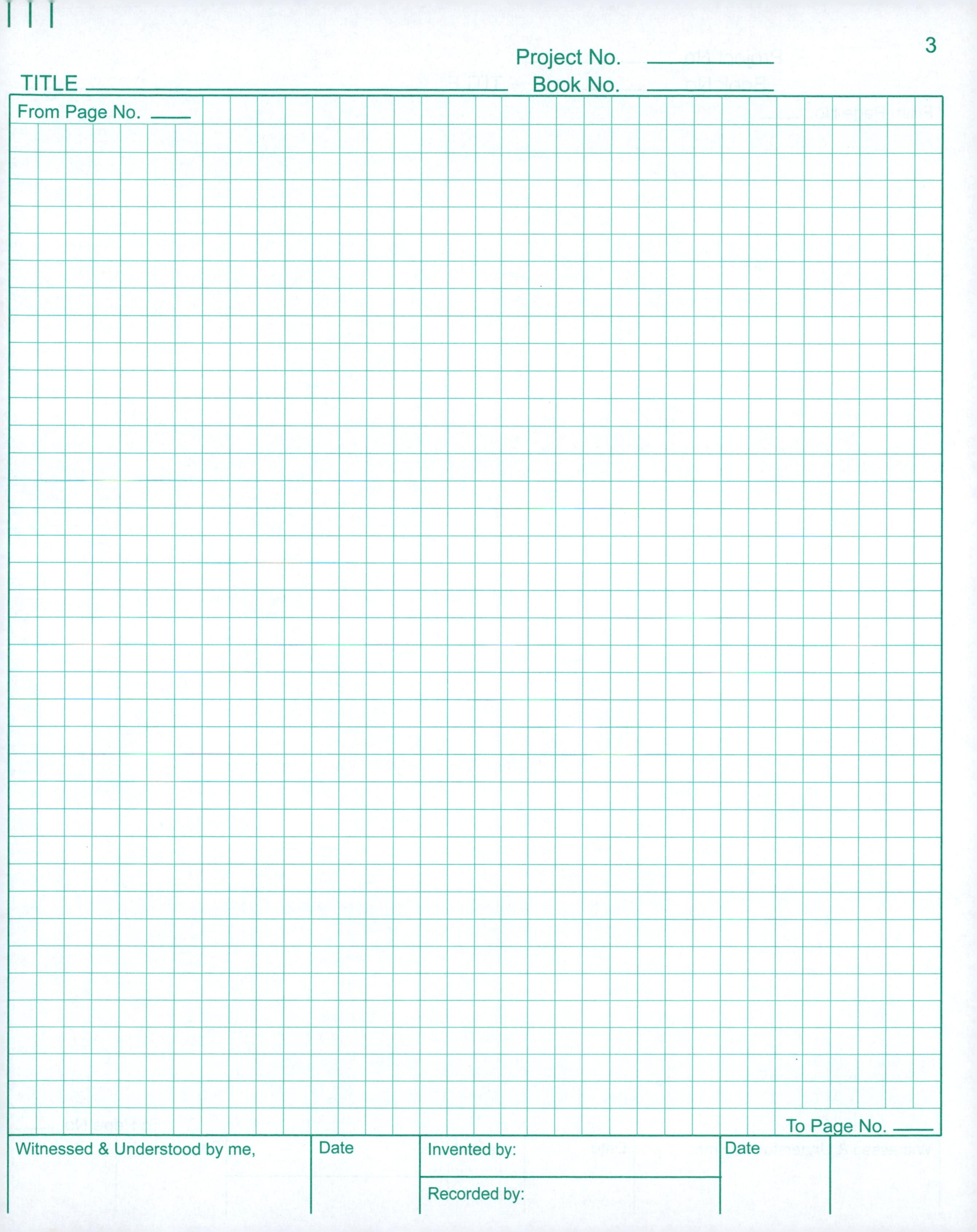
3
Project No.
TITLE
Book No.
From Page No.
To Page No.
Witnessed & Understood by me,
Date
Invented by:
Recorded by:
Date

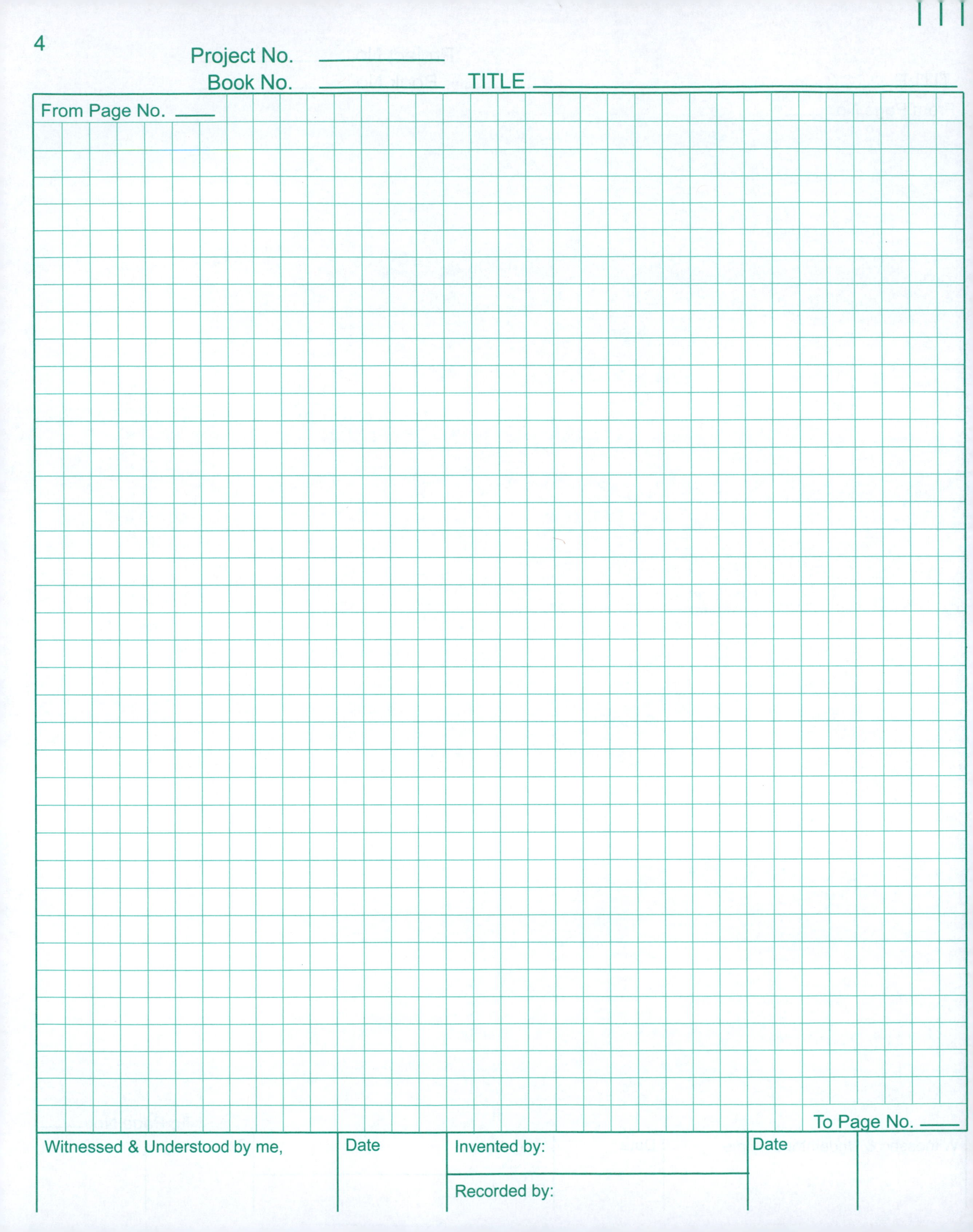
Project No.
Book No.
TITLE
From Page No.
To Page No.
Witnessed & Understood by me,
Date
Invented by:
Recorded by:
Date

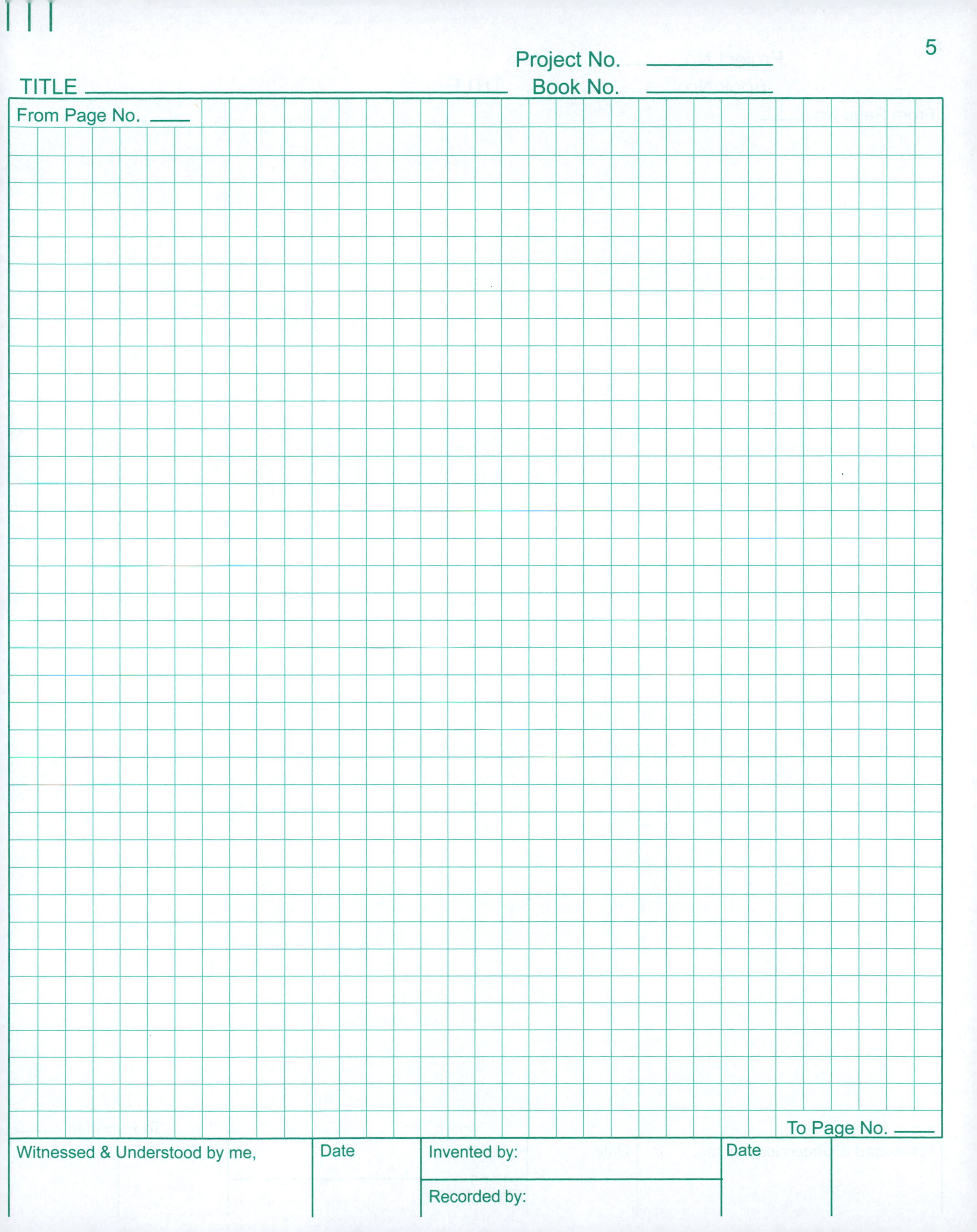

5
Project No.
TITLE
Book No.
From Page No.
To Page No.
Witnessed & Understood by me,
Date
Invented by:
Recorded by:
Date

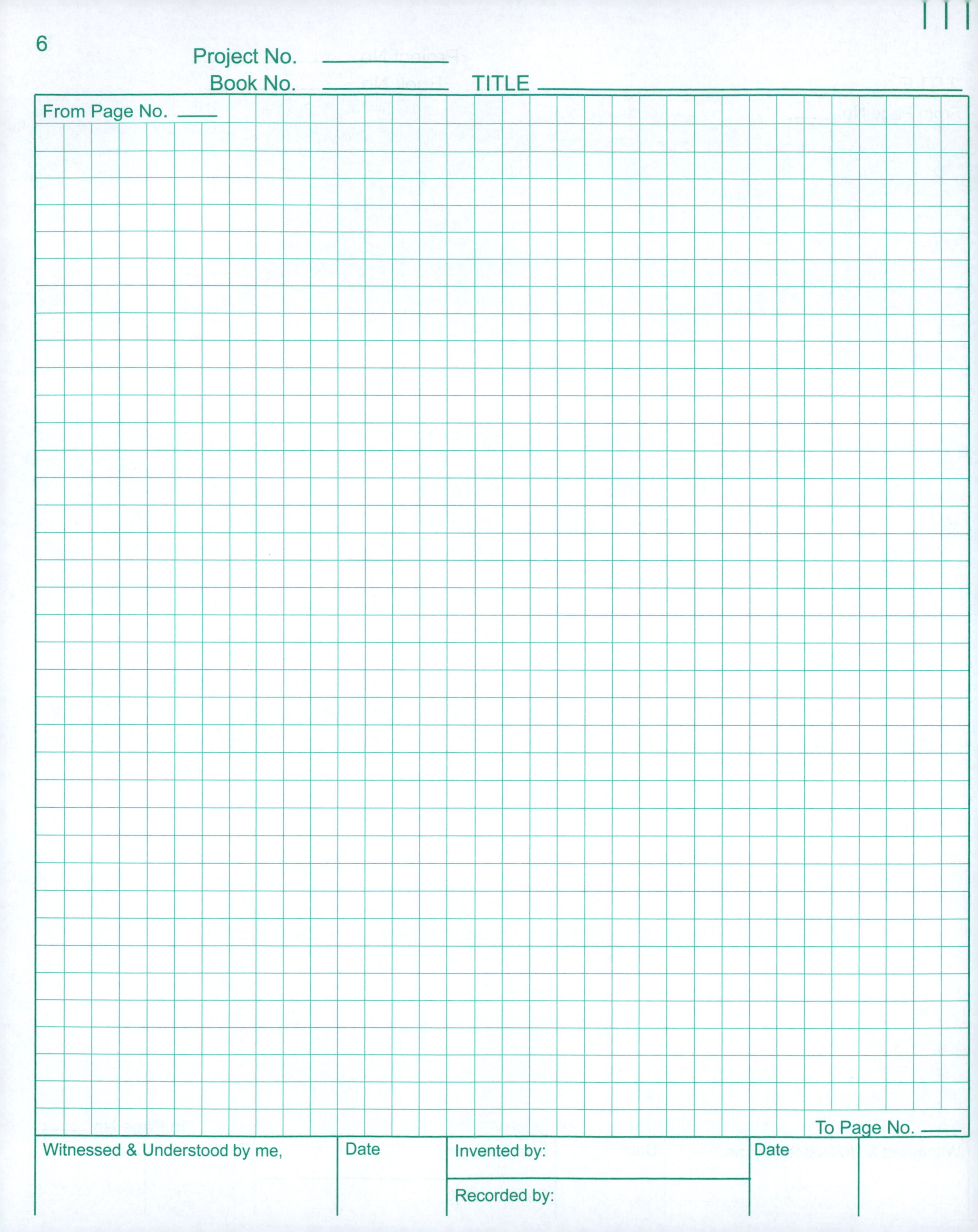
6
Project No.
Book No.
TITLE
From Page No.
To Page No.
Witnessed & Understood by me,
Date
Invented by:
Recorded by:
Date

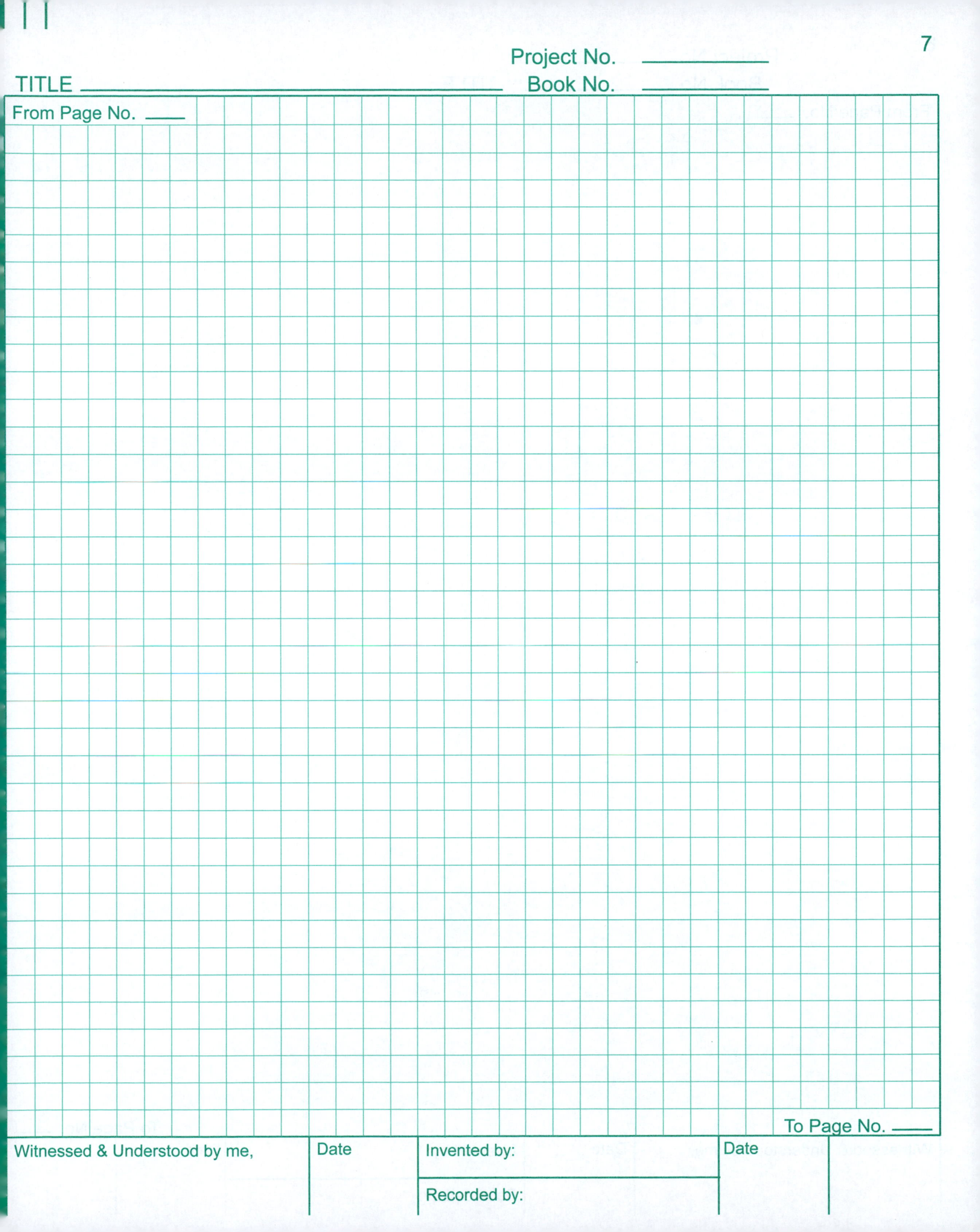

Project No. ______

TITLE ______ Book No. ______

From Page No. ____

To Page No. ____

Witnessed & Understood by me,	Date	Invented by:	Date
		Recorded by:	

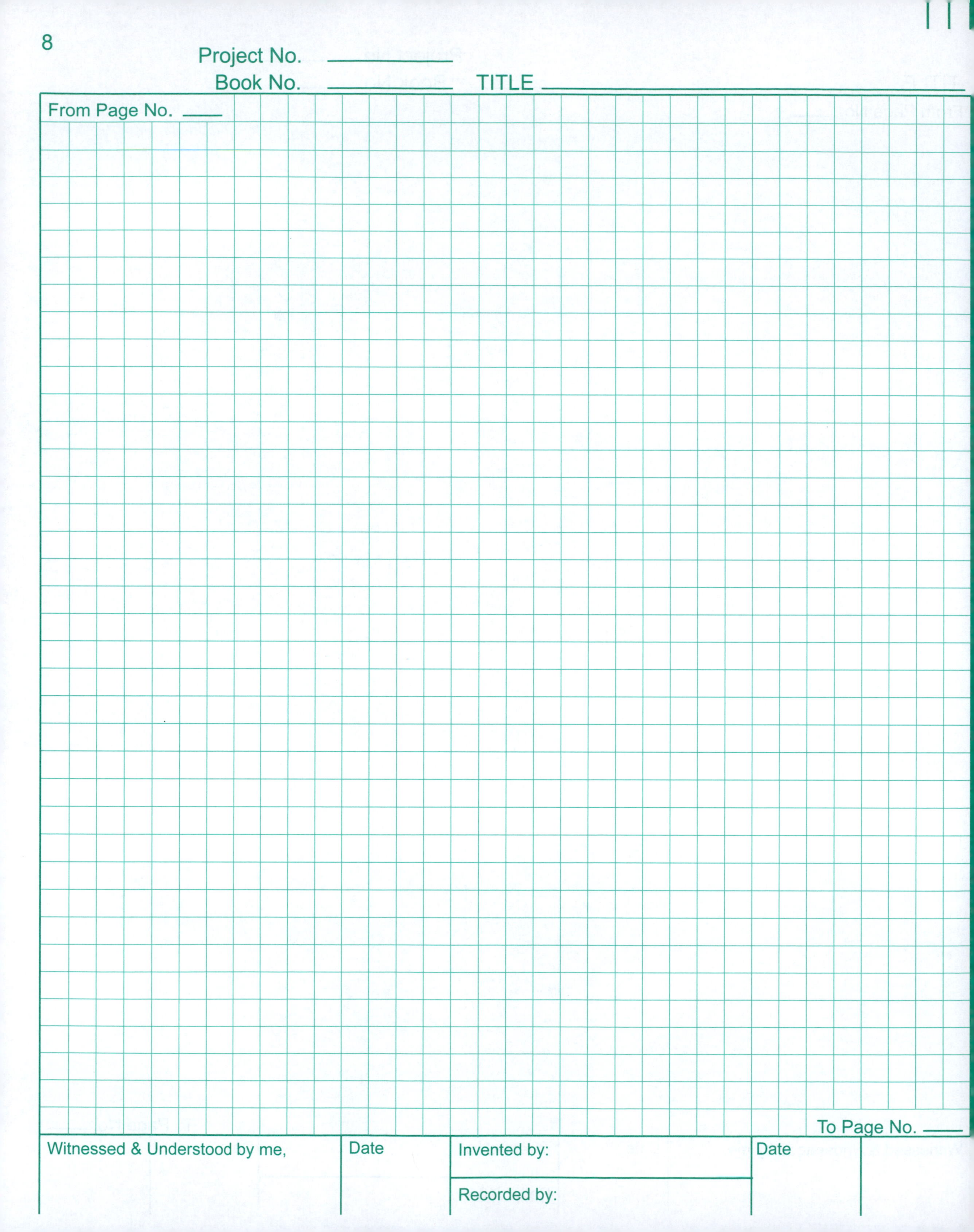
Project No.
Book No.
TITLE
From Page No.
To Page No.
Witnessed & Understood by me,
Date
Invented by:
Recorded by:
Date

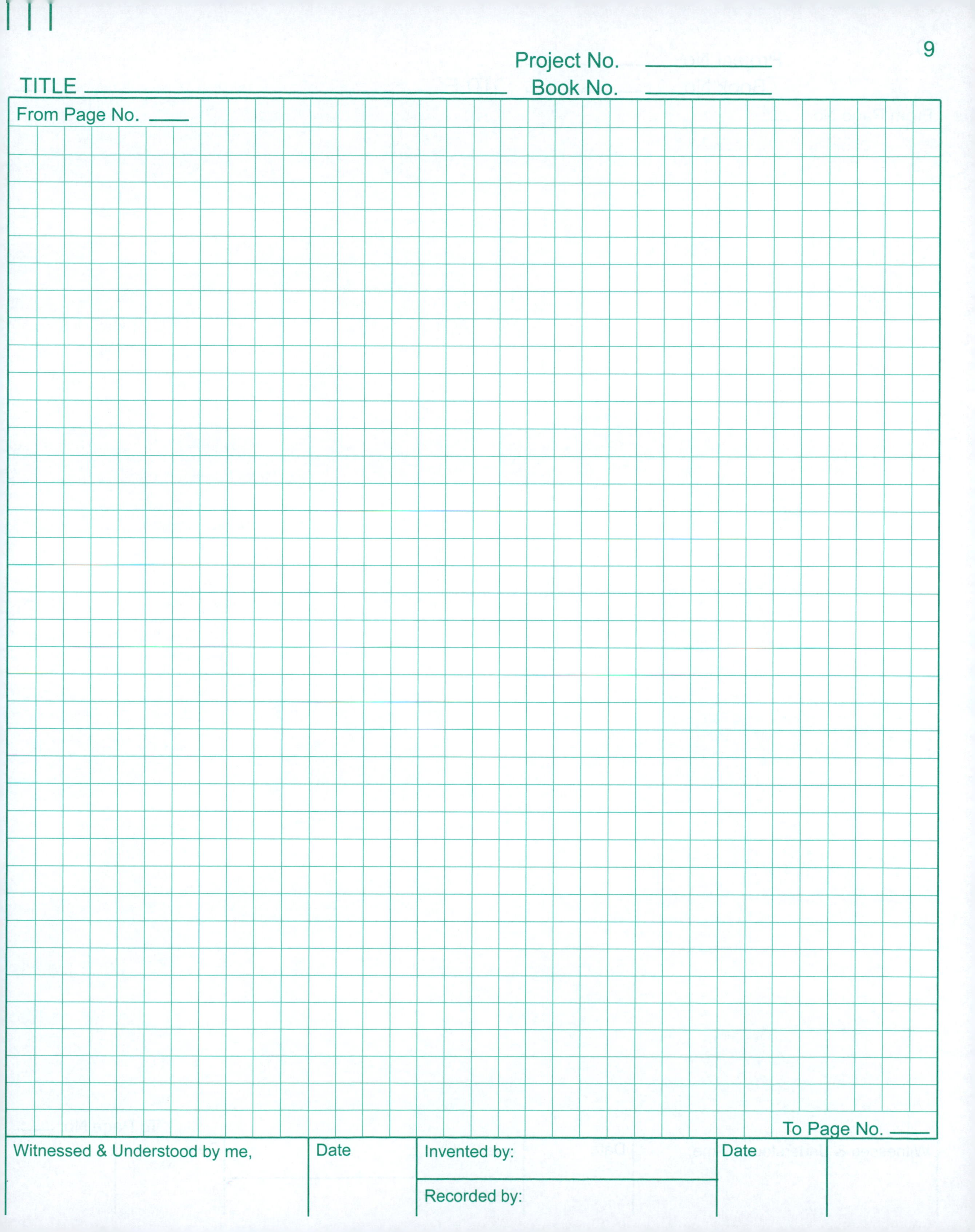

Project No. ____________

TITLE ________________________ Book No. ____________

From Page No. ____

To Page No. ____

Witnessed & Understood by me,	Date	Invented by:	Date
		Recorded by:	

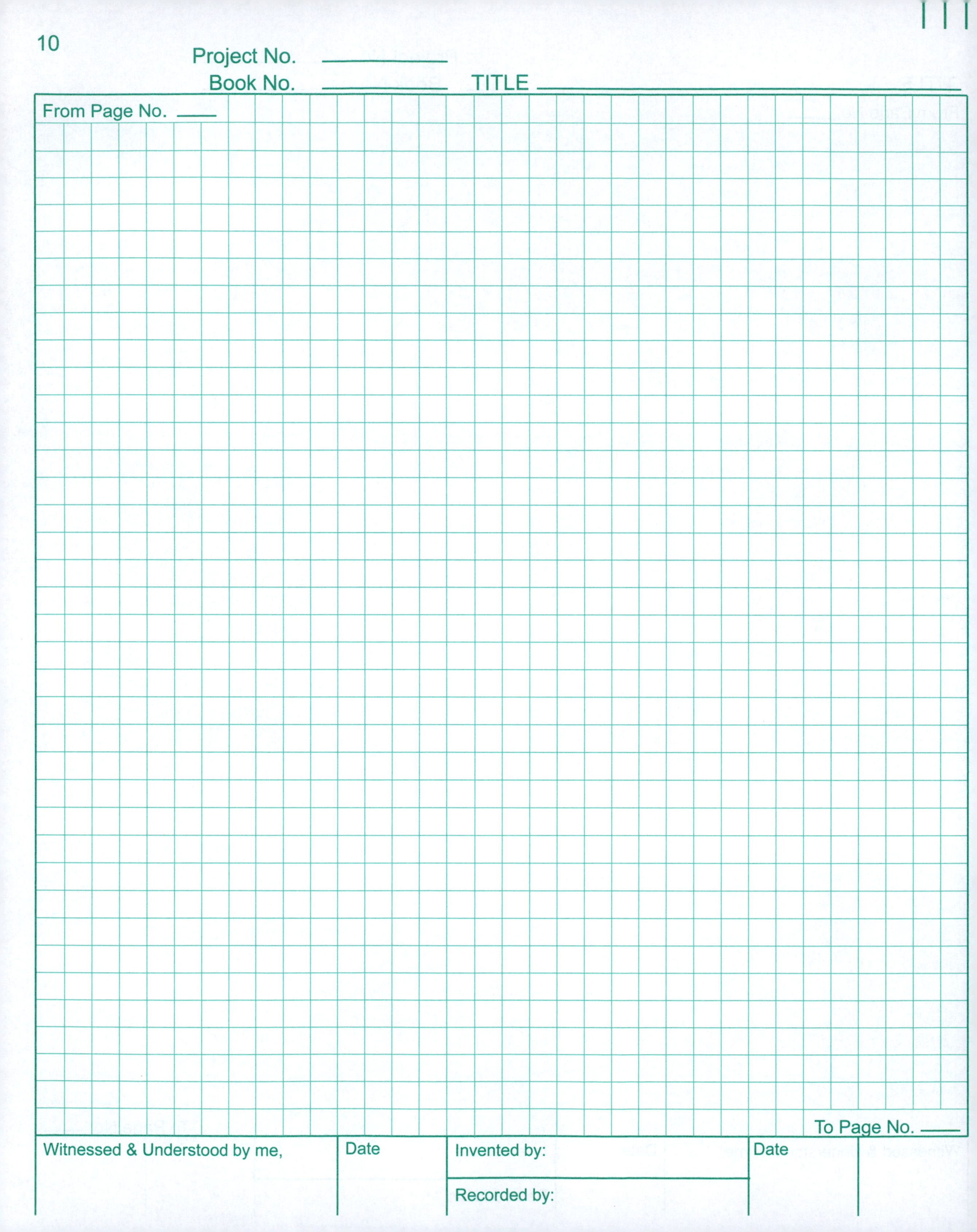
Project No.
Book No.
TITLE
From Page No.
To Page No.
Witnessed & Understood by me,
Date
Invented by:
Recorded by:
Date

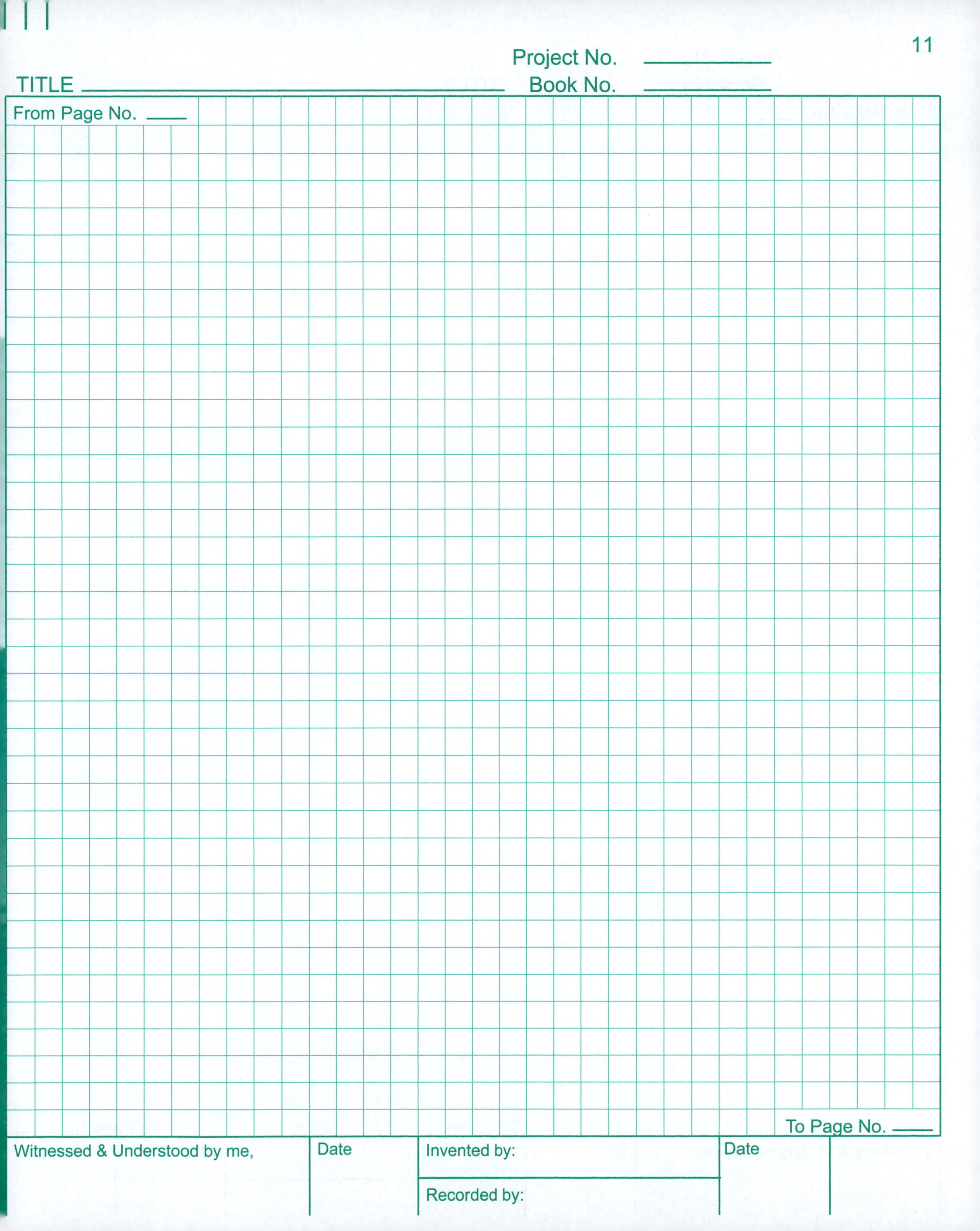

Project No. ____________

TITLE ____ Book No. ____________

From Page No. ____

To Page No. ____

Witnessed & Understood by me,	Date	Invented by:	Date
		Recorded by:	

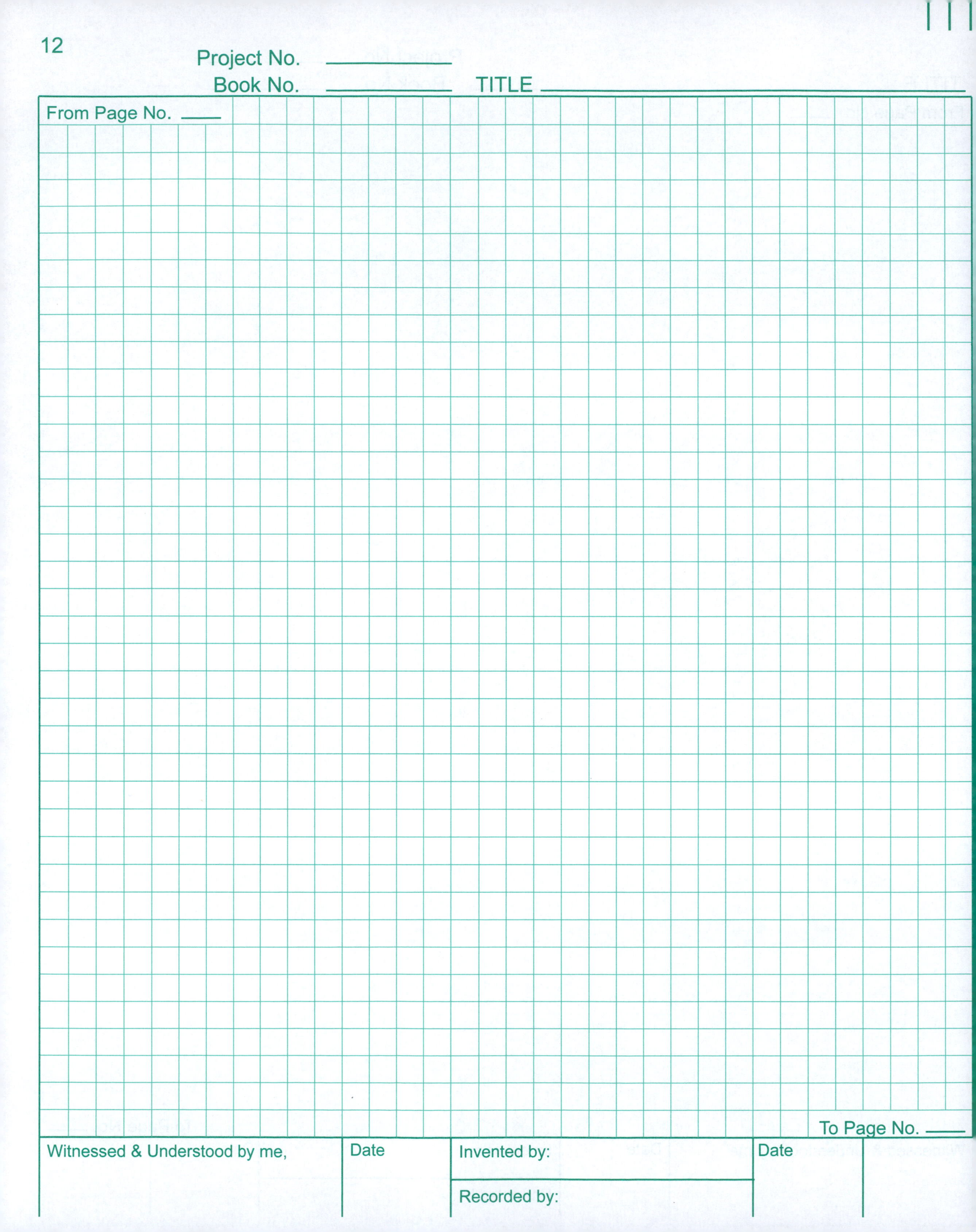
Project No.
Book No.
TITLE
From Page No.
To Page No.
Witnessed & Understood by me,
Date
Invented by:
Recorded by:
Date

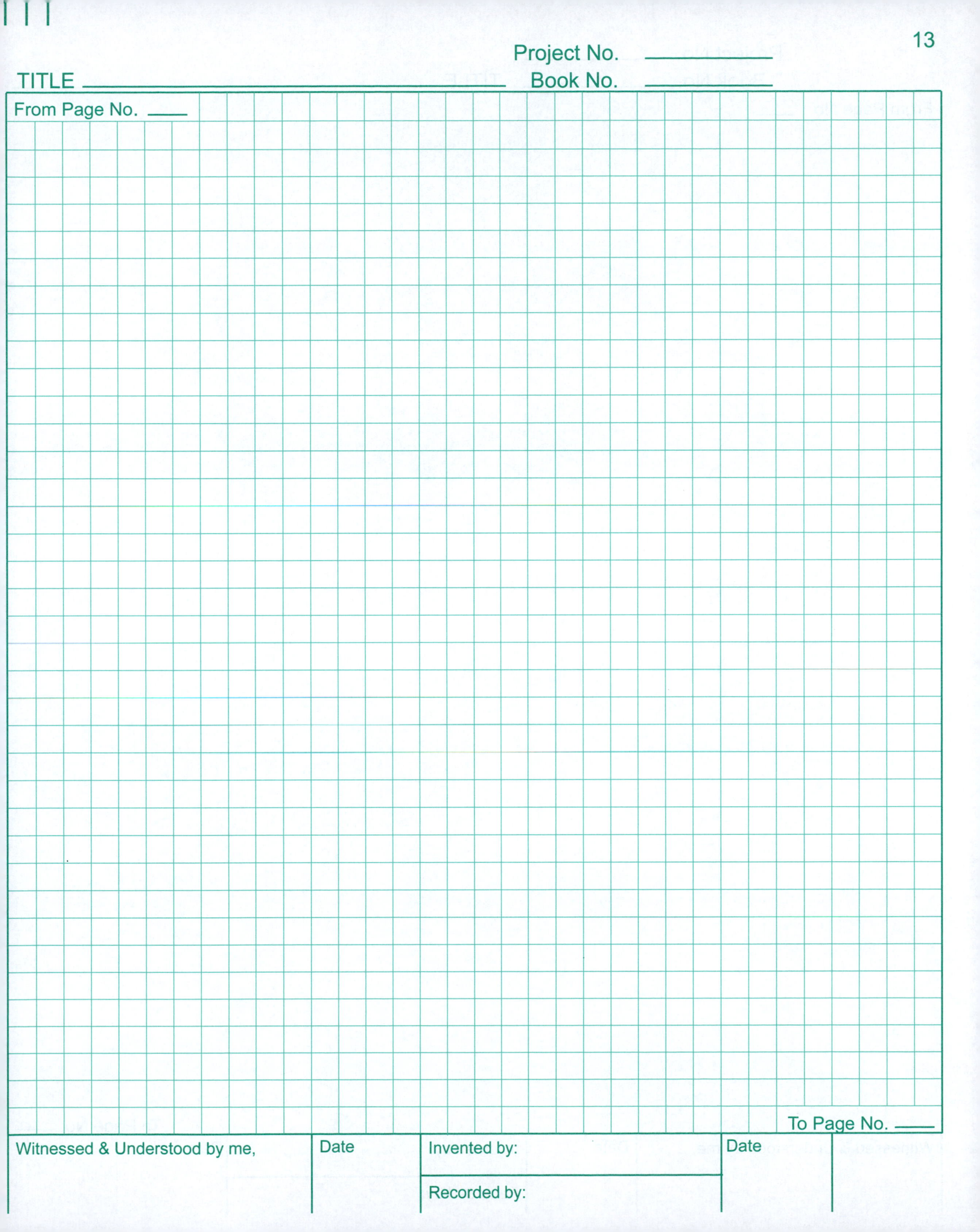
Project No.
TITLE
Book No.
From Page No.
To Page No.
Witnessed & Understood by me,
Date
Invented by:
Recorded by:
Date

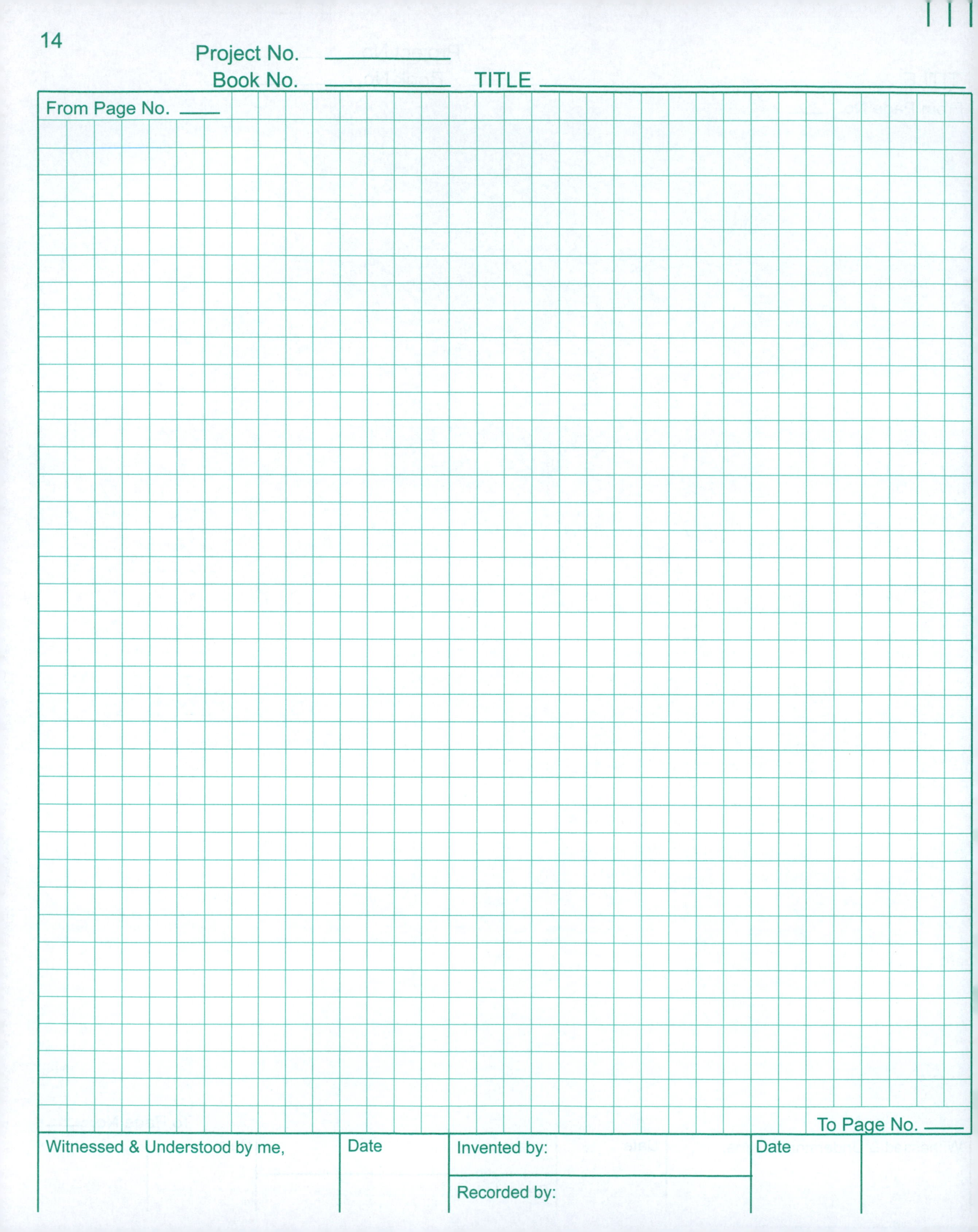
Project No.
Book No.
TITLE
From Page No.
To Page No.
Witnessed & Understood by me,
Date
Invented by:
Recorded by:
Date

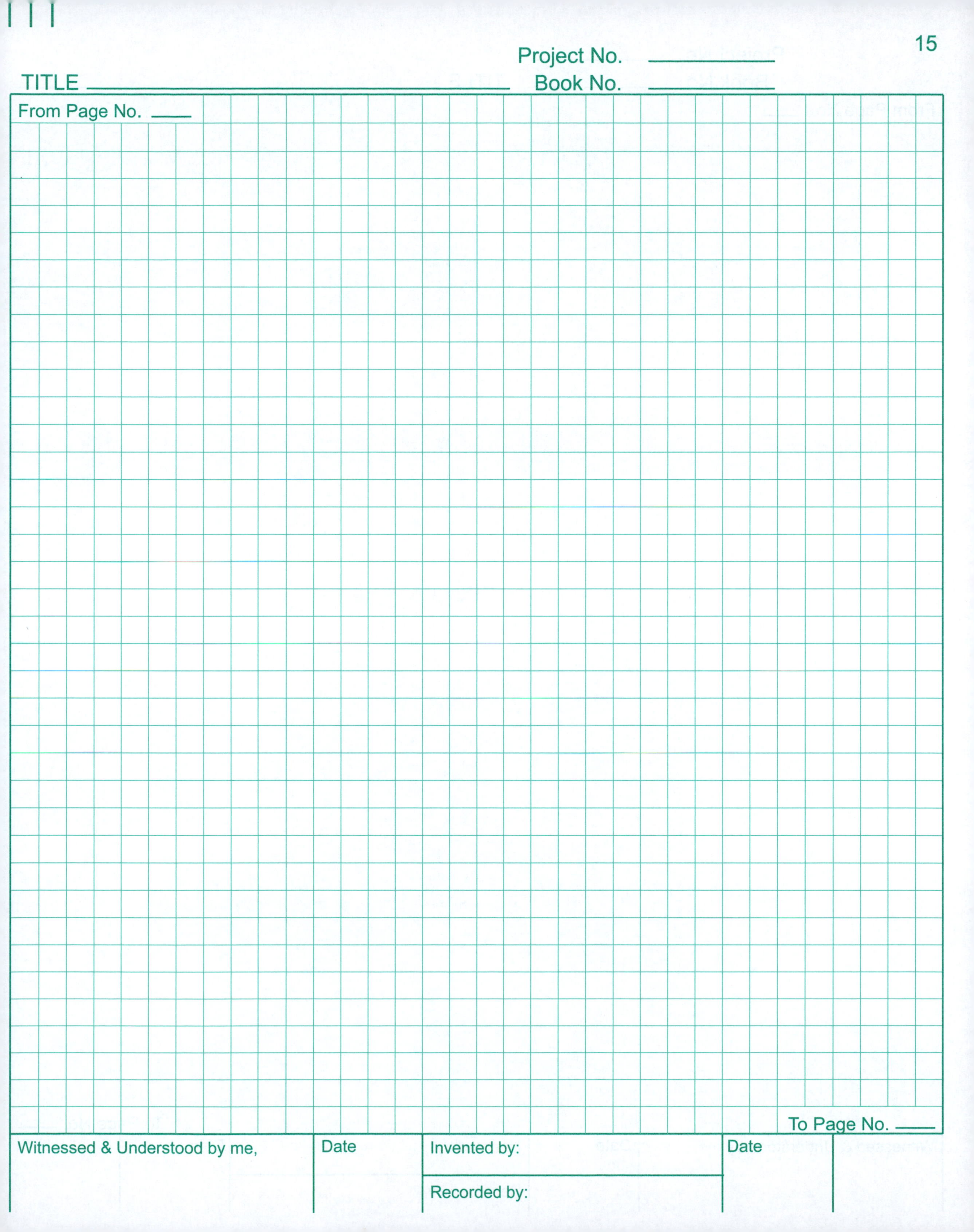
Project No.
TITLE
Book No.
From Page No.
To Page No.
Witnessed & Understood by me,
Date
Invented by:
Recorded by:
Date

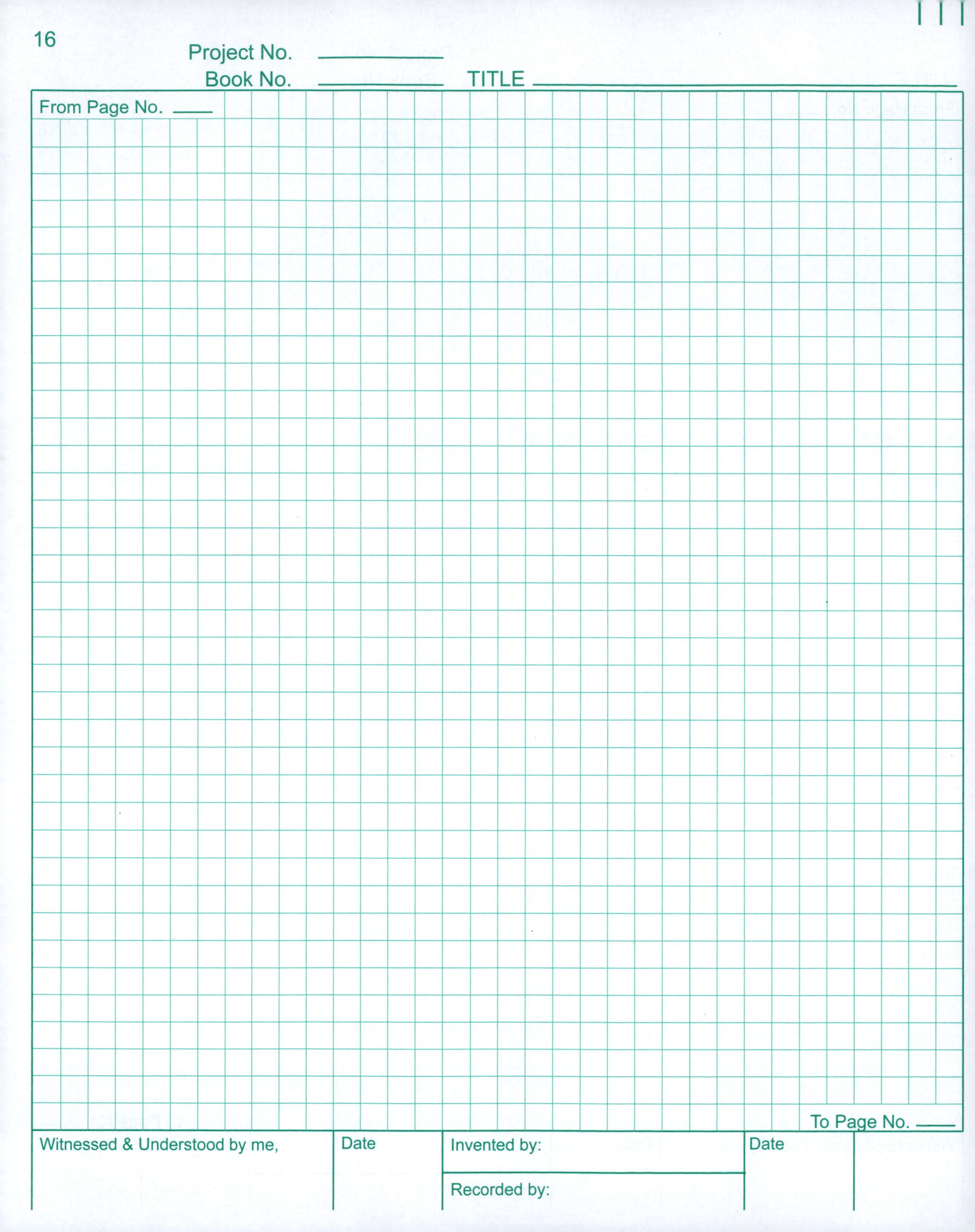
Project No.
Book No.
TITLE
From Page No.
To Page No.
Witnessed & Understood by me,
Date
Invented by:
Recorded by:
Date

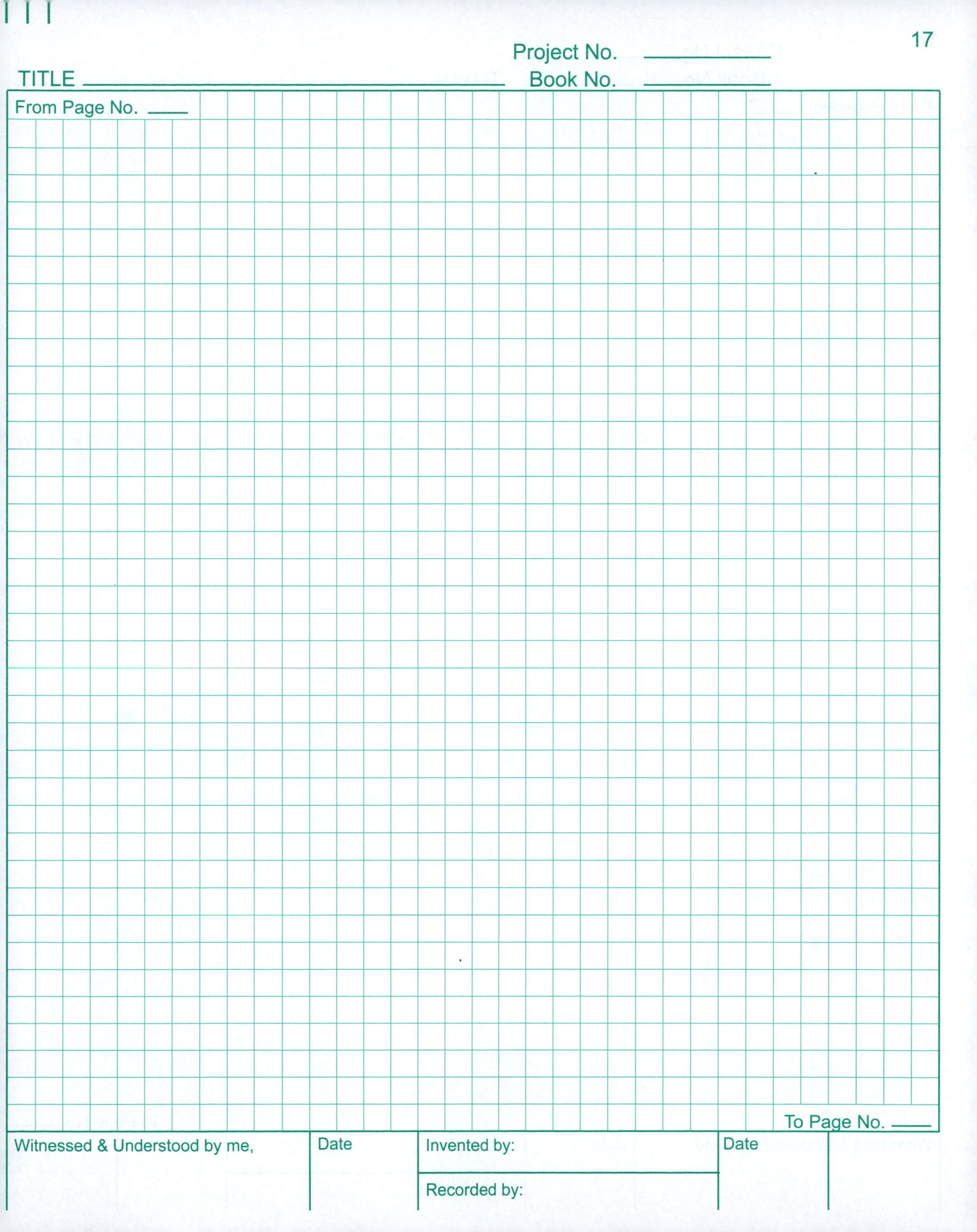
Project No.
TITLE
Book No.
From Page No.
To Page No.
Witnessed & Understood by me,
Date
Invented by:
Recorded by:
Date

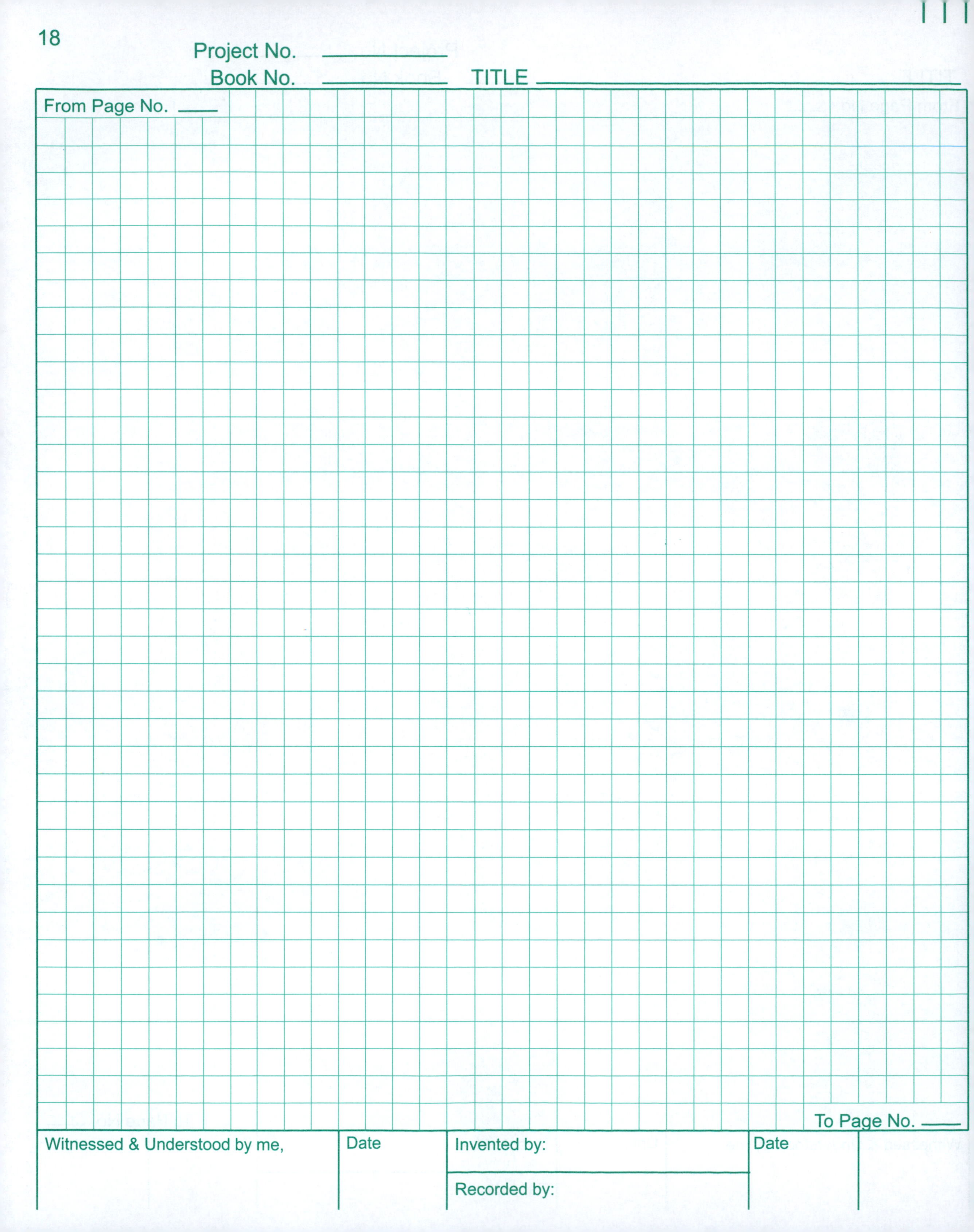
Project No.
Book No.
TITLE
From Page No.
To Page No.
Witnessed & Understood by me,
Date
Invented by:
Recorded by:
Date

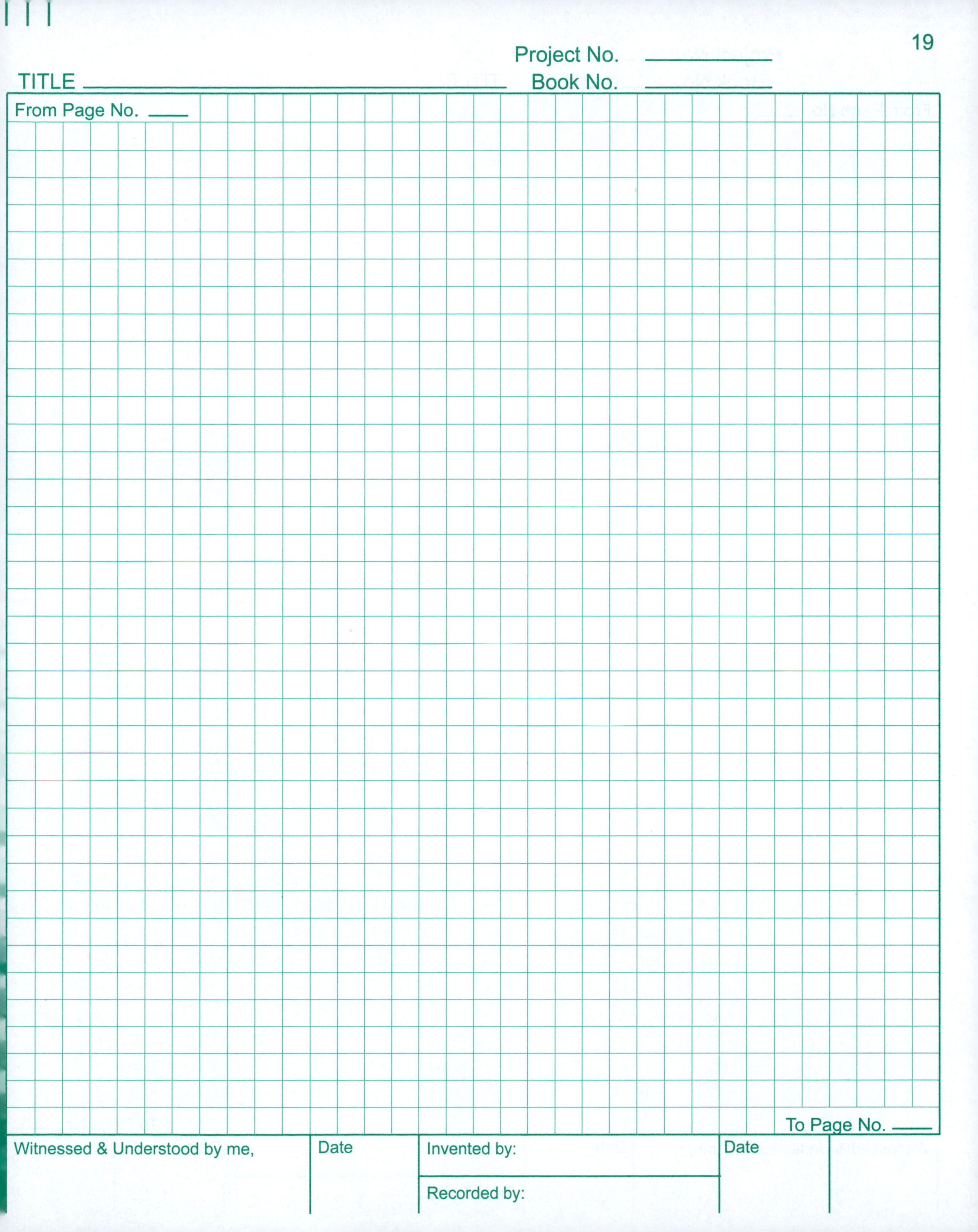

19
Project No.
TITLE
Book No.
From Page No.
To Page No.
Witnessed & Understood by me,
Date
Invented by:
Recorded by:
Date

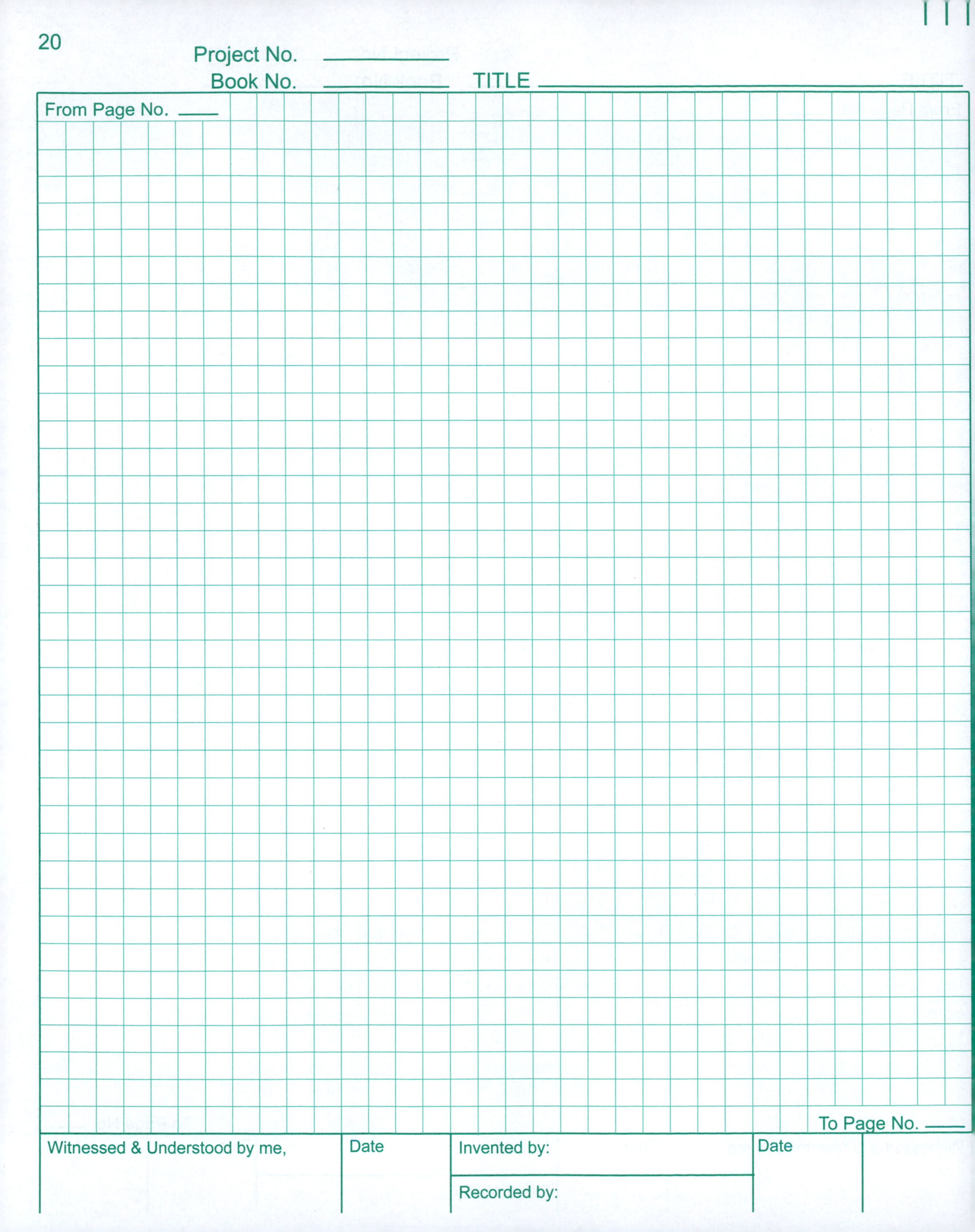

Project No.

Book No. TITLE

From Page No.

To Page No.

Witnessed & Understood by me,	Date	Invented by:	Date
		Recorded by:	

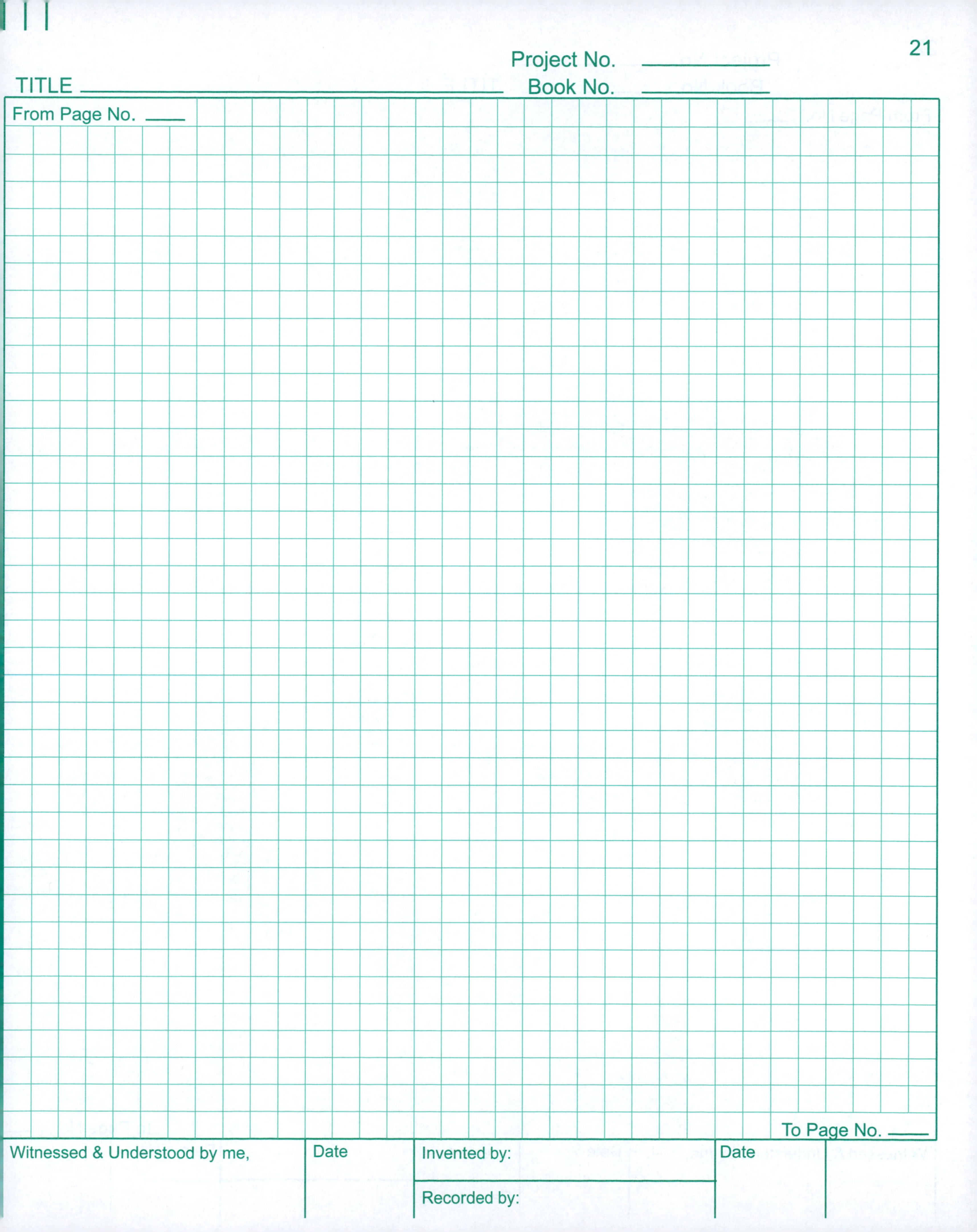
Project No.
TITLE
Book No.
From Page No.
To Page No.
Witnessed & Understood by me,
Date
Invented by:
Date
Recorded by:

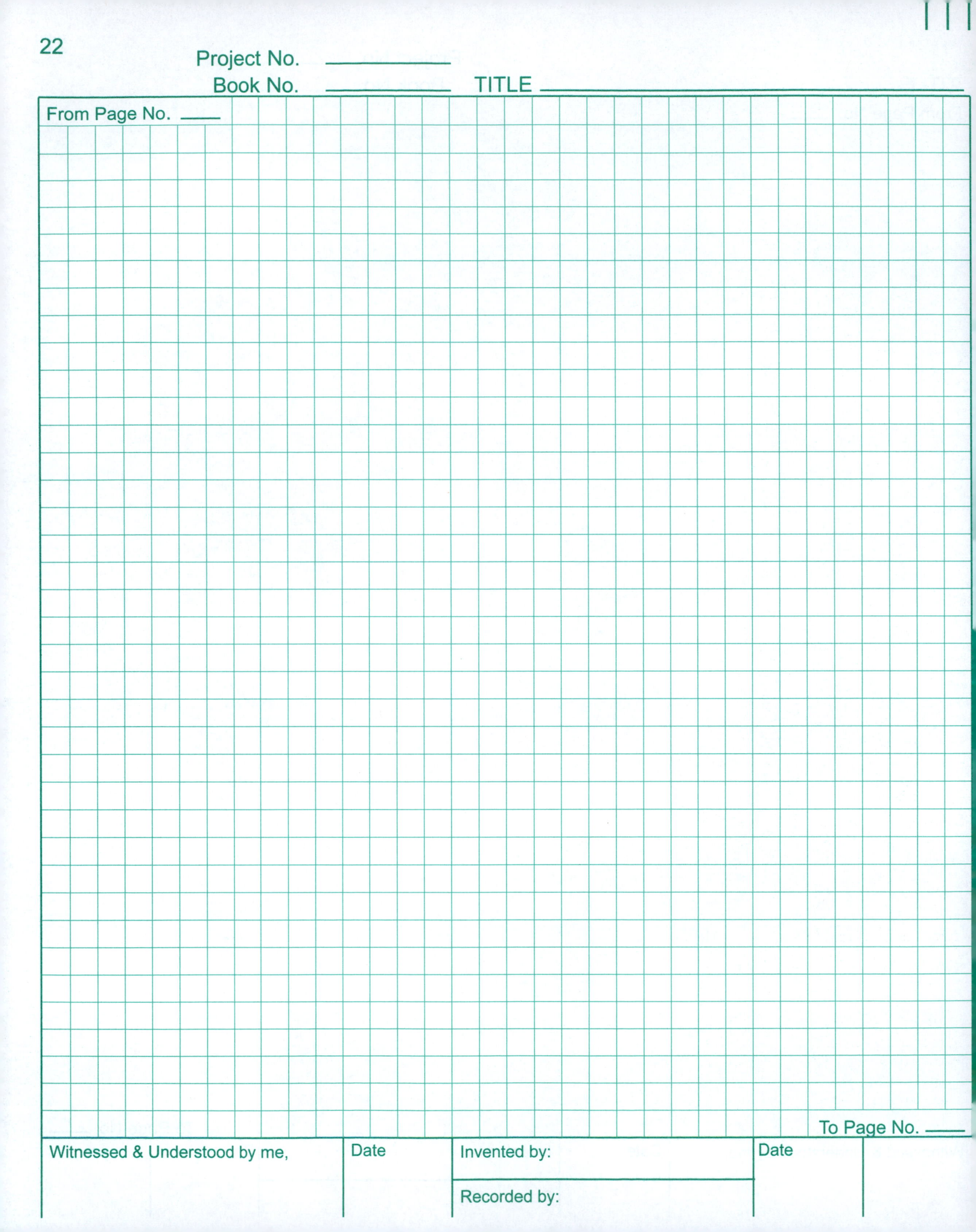

Project No. ____________

Book No. ____________ TITLE ____________

From Page No. ____

To Page No. ____

Witnessed & Understood by me,	Date	Invented by:	Date
		Recorded by:	

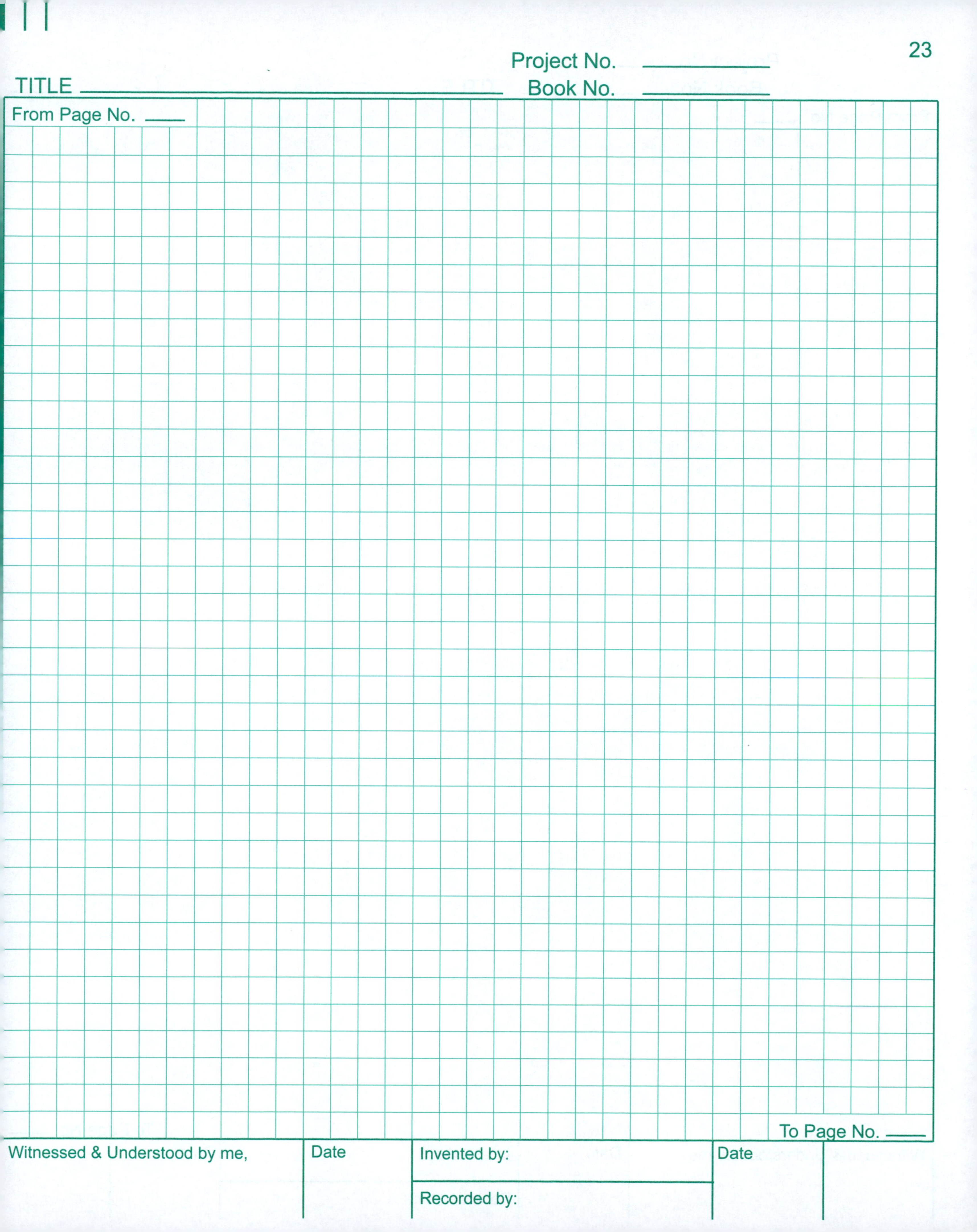

TITLE ________ Project No. ________ Book No. ________

From Page No. ____

To Page No. ____

Witnessed & Understood by me,	Date	Invented by:	Date
		Recorded by:	

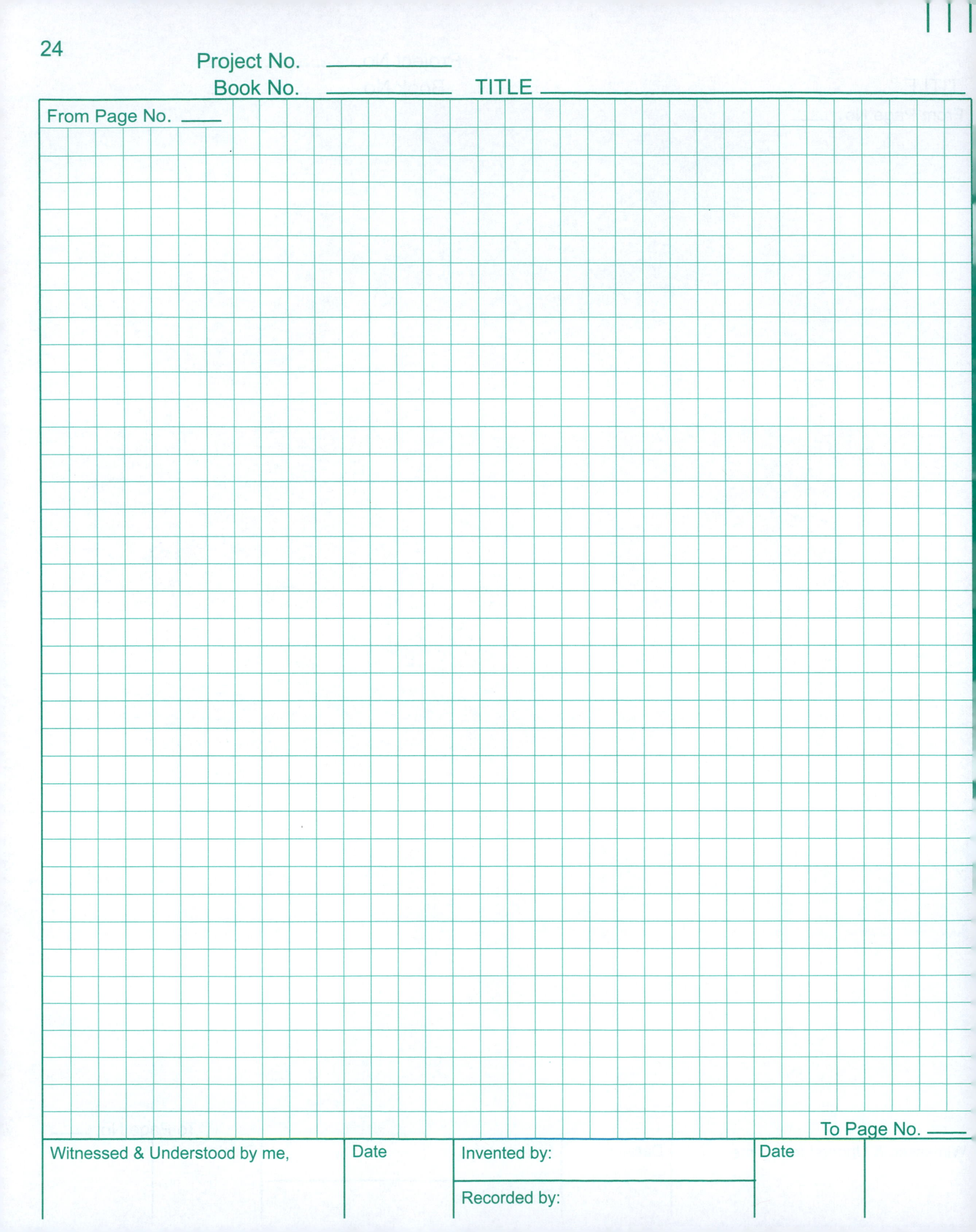
Project No.
Book No.
TITLE
From Page No.
To Page No.
Witnessed & Understood by me,
Date
Invented by:
Recorded by:
Date

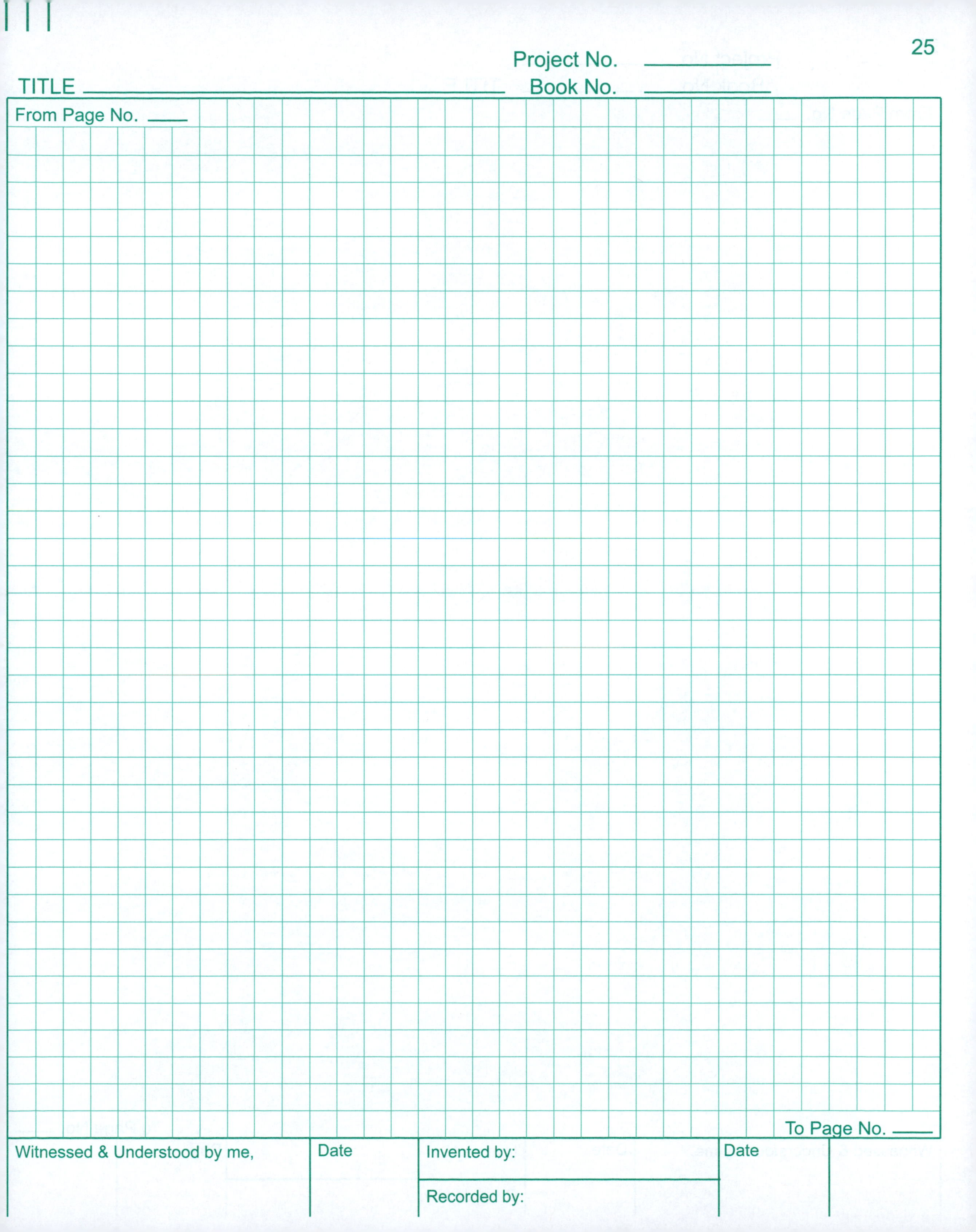
Project No.
TITLE
Book No.
From Page No.
To Page No.
Witnessed & Understood by me,
Date
Invented by:
Recorded by:
Date

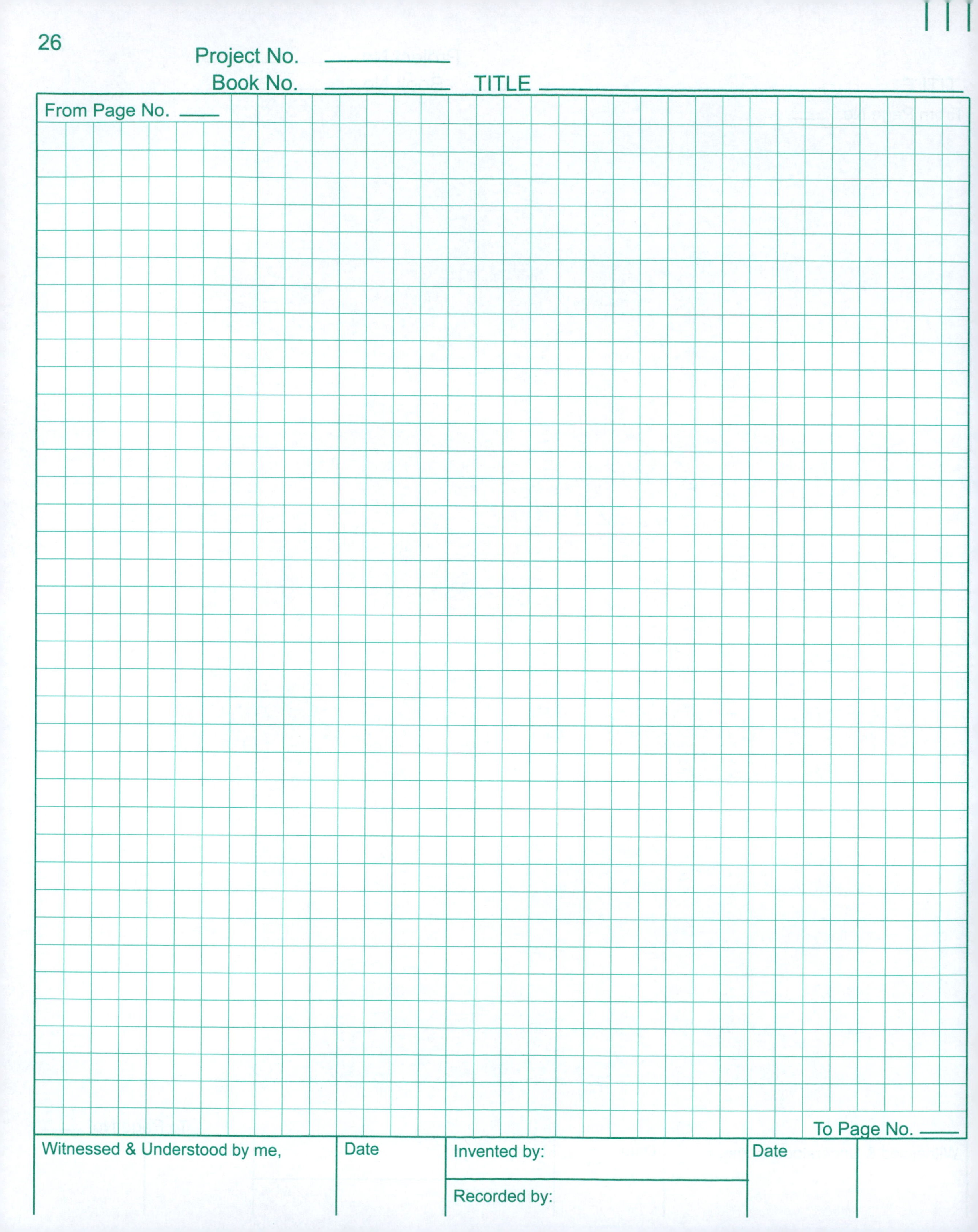

Project No. ____________

Book No. ____________ TITLE ______________________________

From Page No. ____

To Page No. ____

Witnessed & Understood by me,	Date	Invented by:	Date
		Recorded by:	

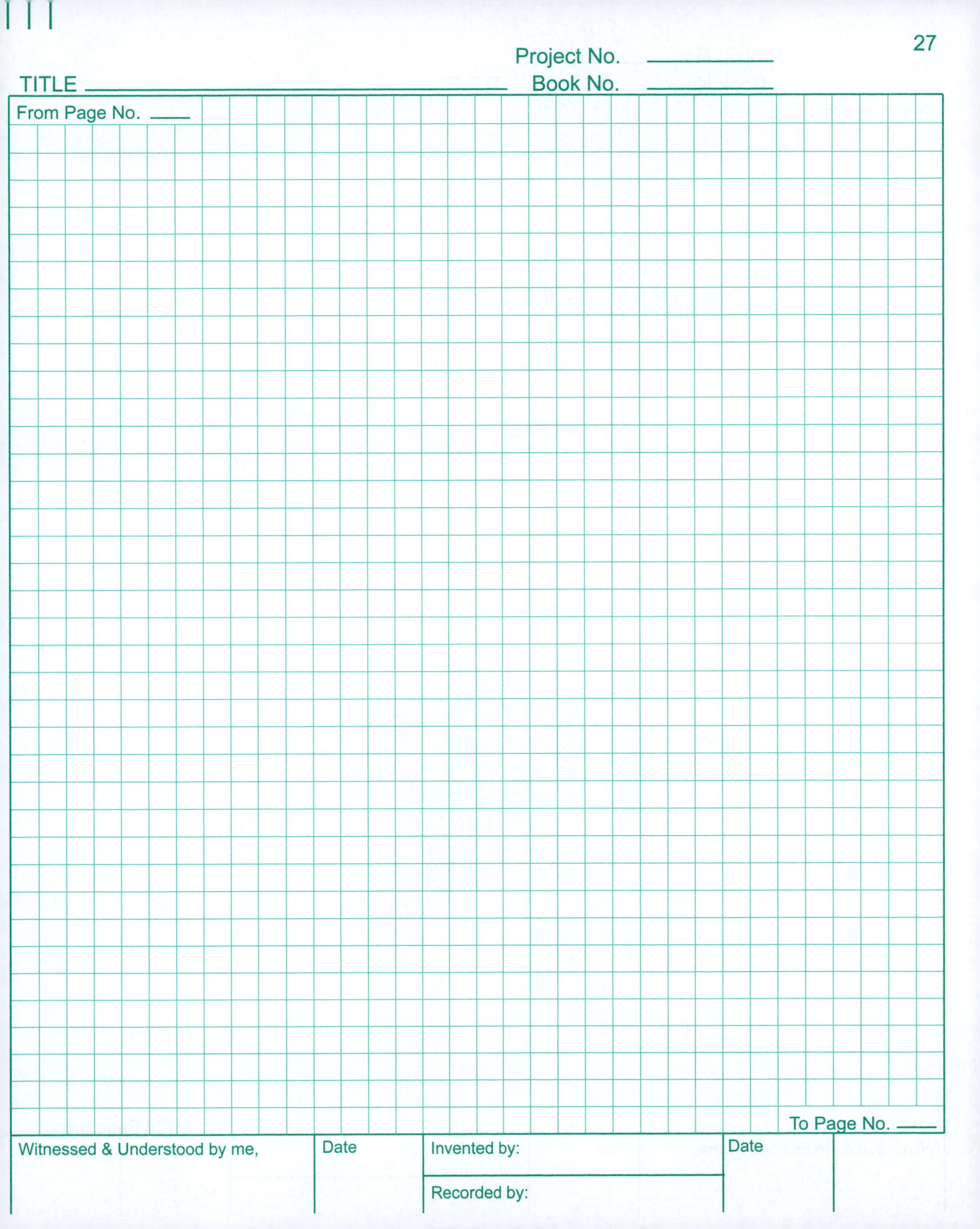
Project No.
TITLE
Book No.
From Page No.
To Page No.
Witnessed & Understood by me,
Date
Invented by:
Recorded by:
Date

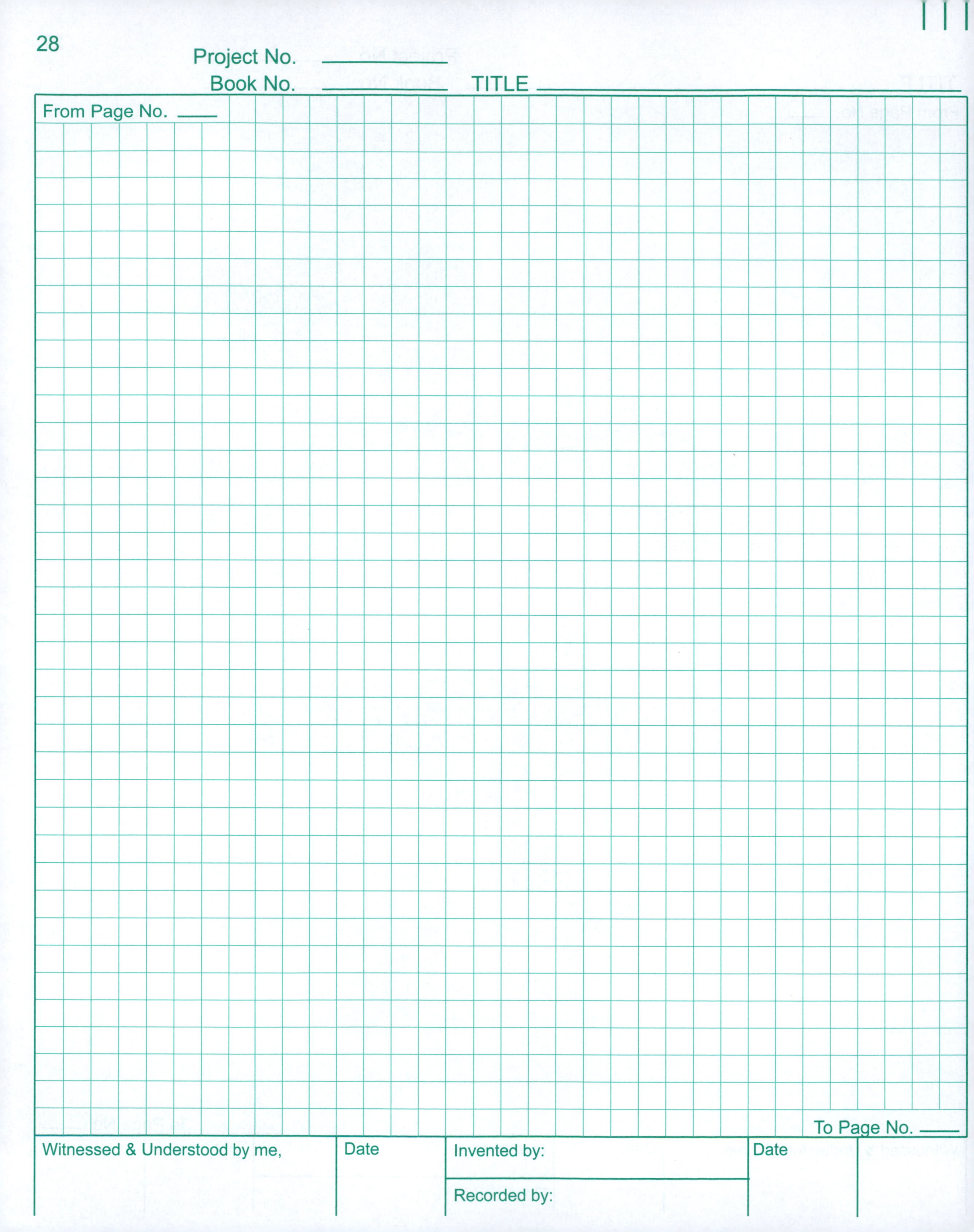
Project No.
Book No.
TITLE
From Page No.
To Page No.
Witnessed & Understood by me,
Date
Invented by:
Recorded by:
Date

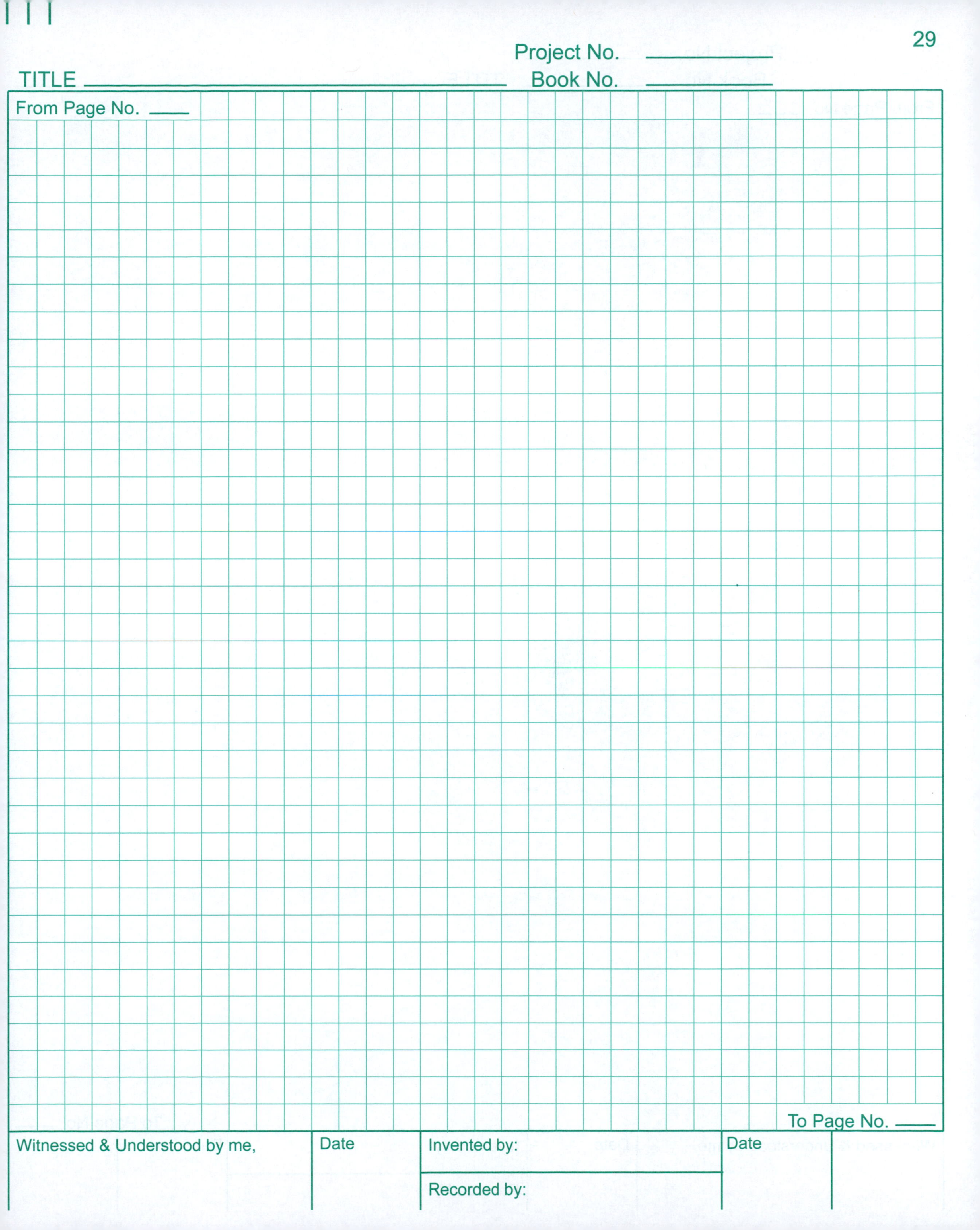
Project No.
TITLE
Book No.
From Page No.
To Page No.
Witnessed & Understood by me,
Date
Invented by:
Recorded by:
Date

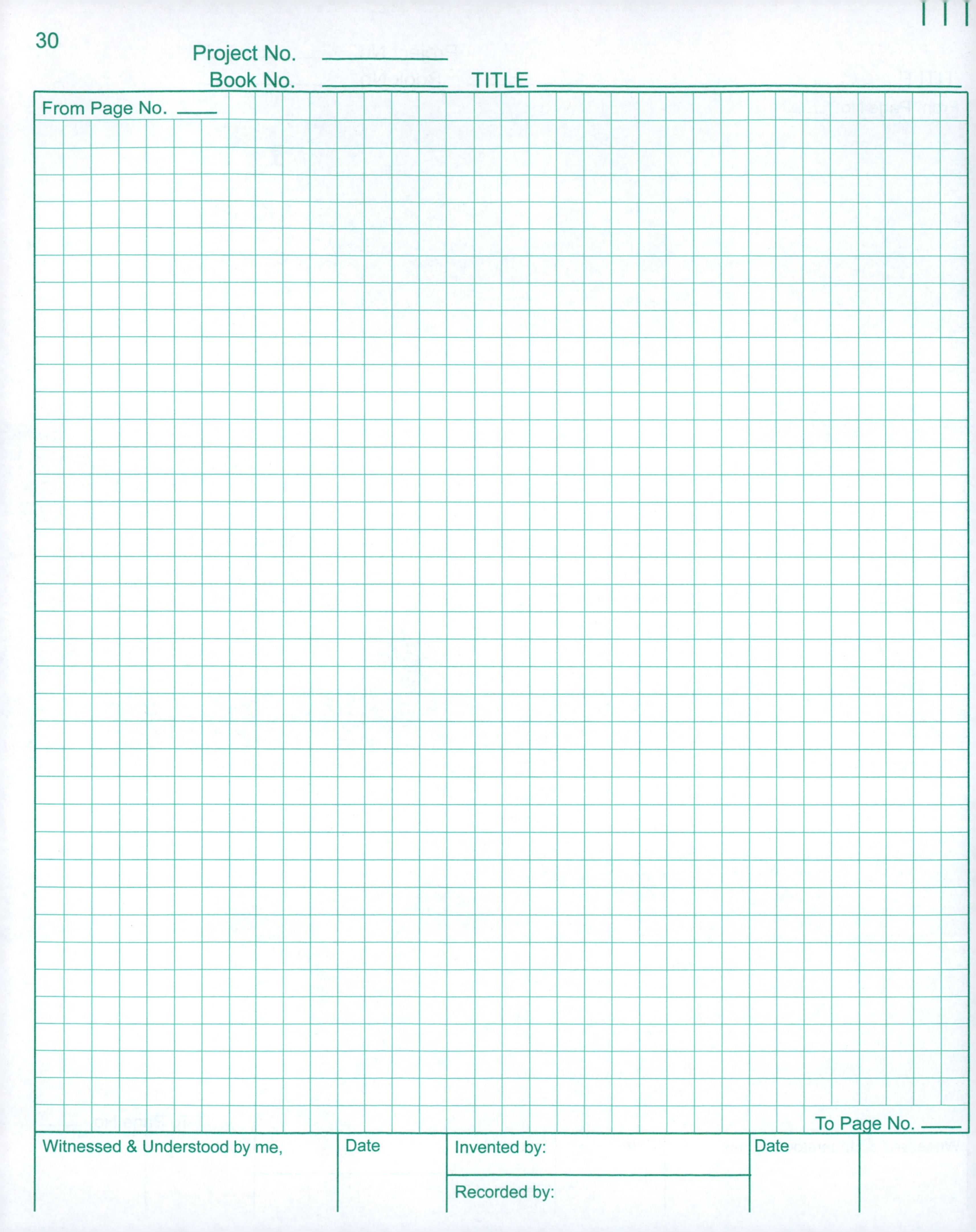

Project No.
Book No.
TITLE
From Page No.
To Page No.
Witnessed & Understood by me,
Date
Invented by:
Recorded by:
Date

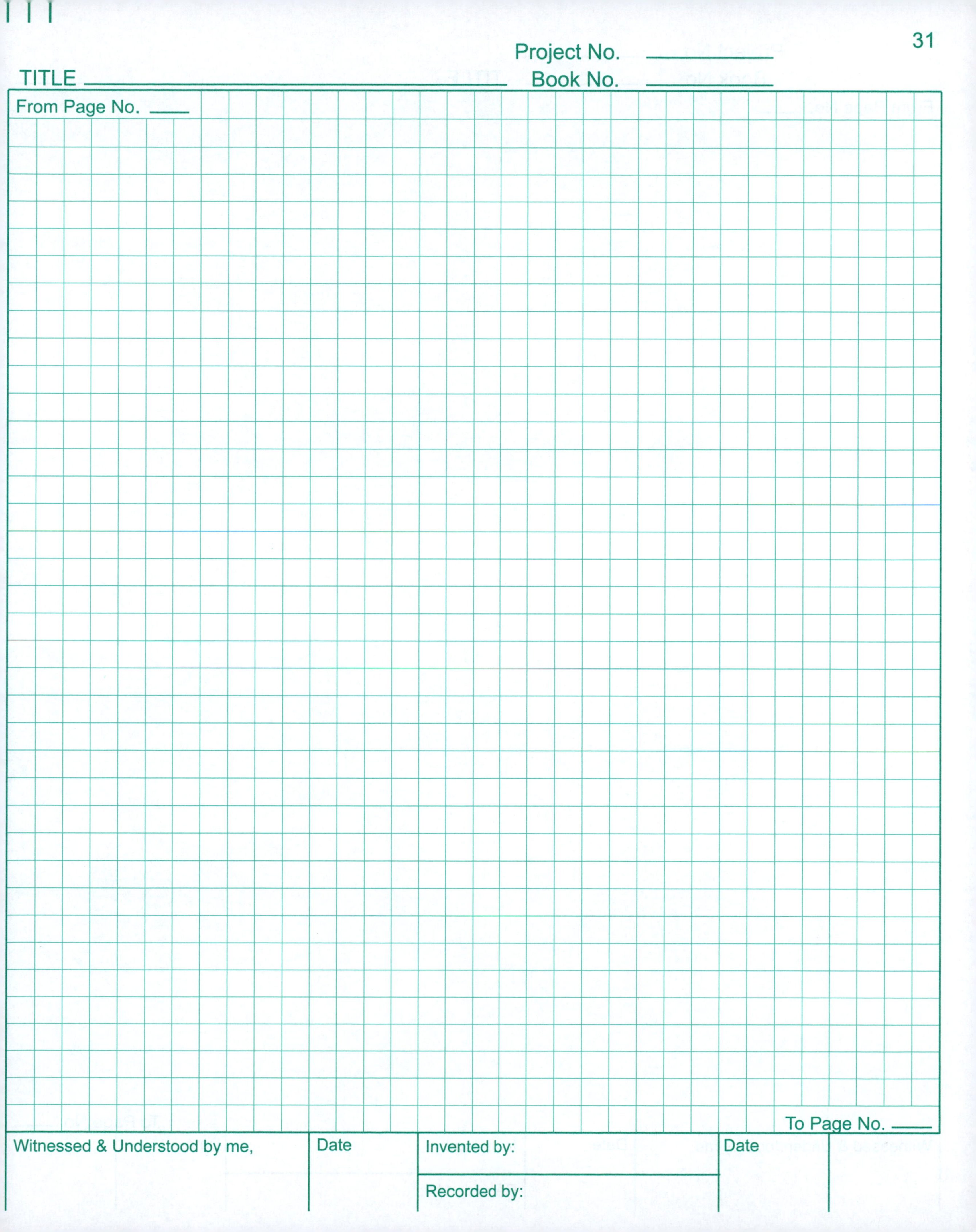

Project No. ____________

TITLE ____________________ Book No. ____________

From Page No. ____

To Page No. ____

Witnessed & Understood by me,	Date	Invented by:	Date
		Recorded by:	

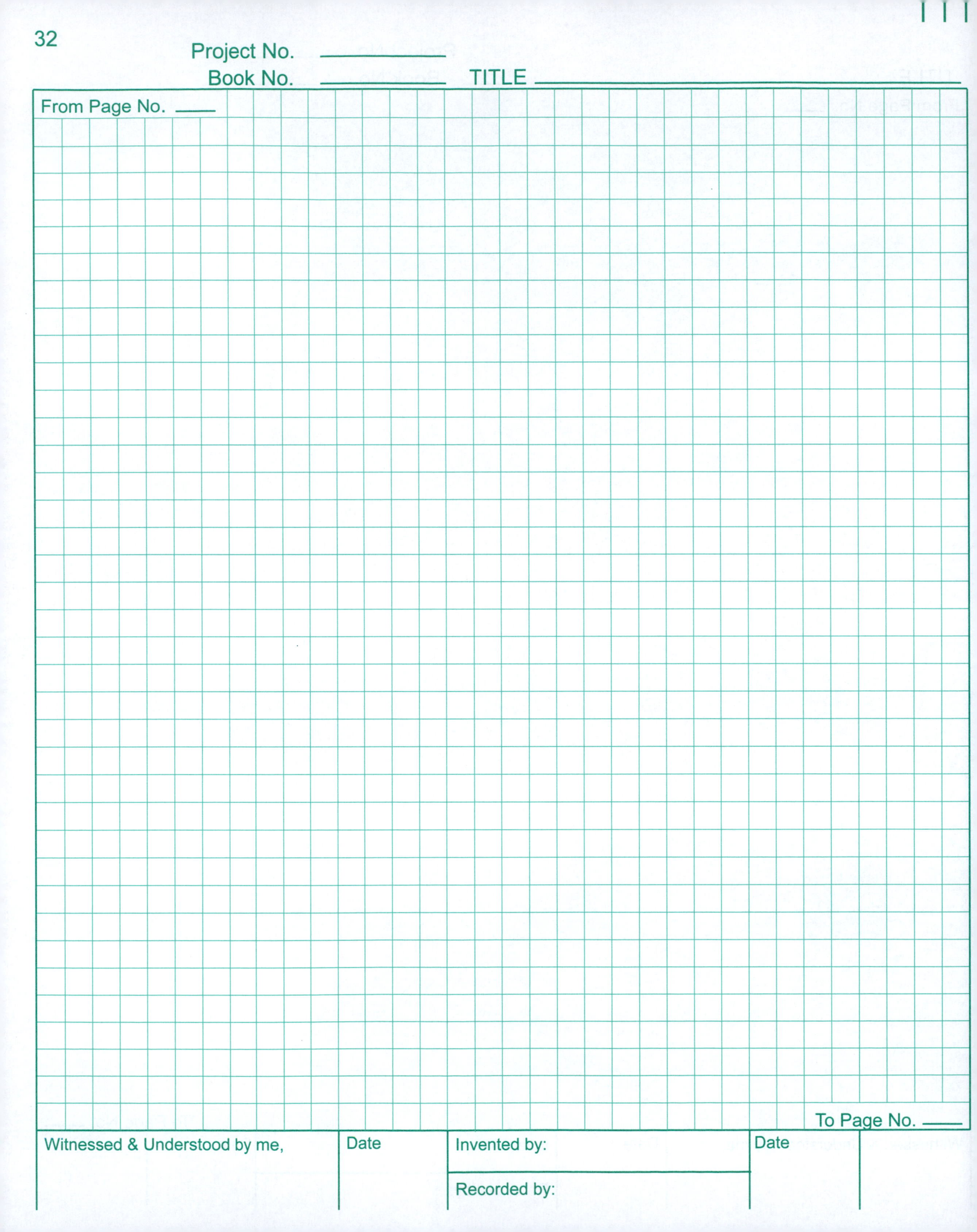
Project No.
Book No.
TITLE
From Page No.
To Page No.
Witnessed & Understood by me,
Date
Invented by:
Recorded by:
Date

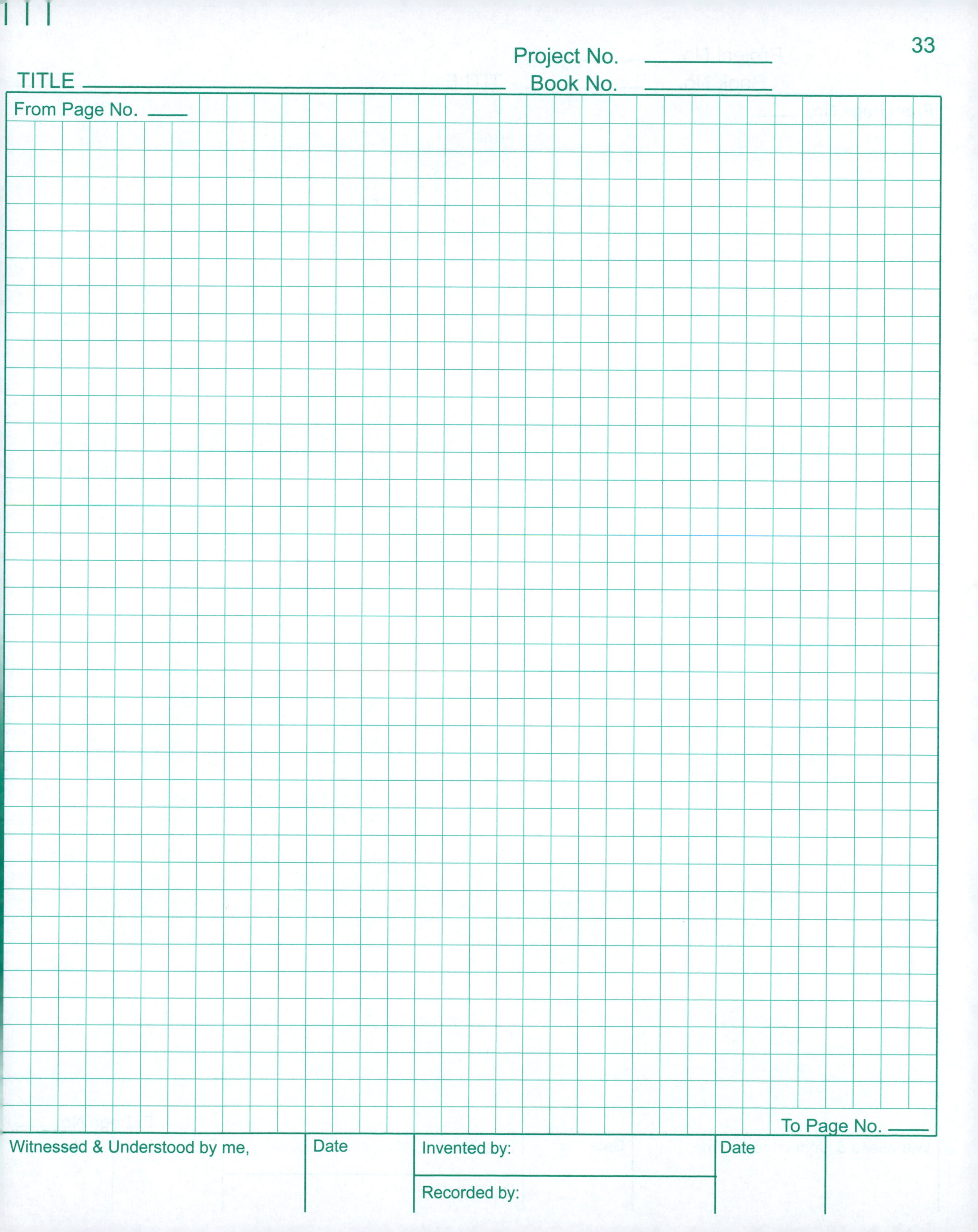
Project No.
TITLE
Book No.
From Page No.
To Page No.
Witnessed & Understood by me,
Date
Invented by:
Date
Recorded by:

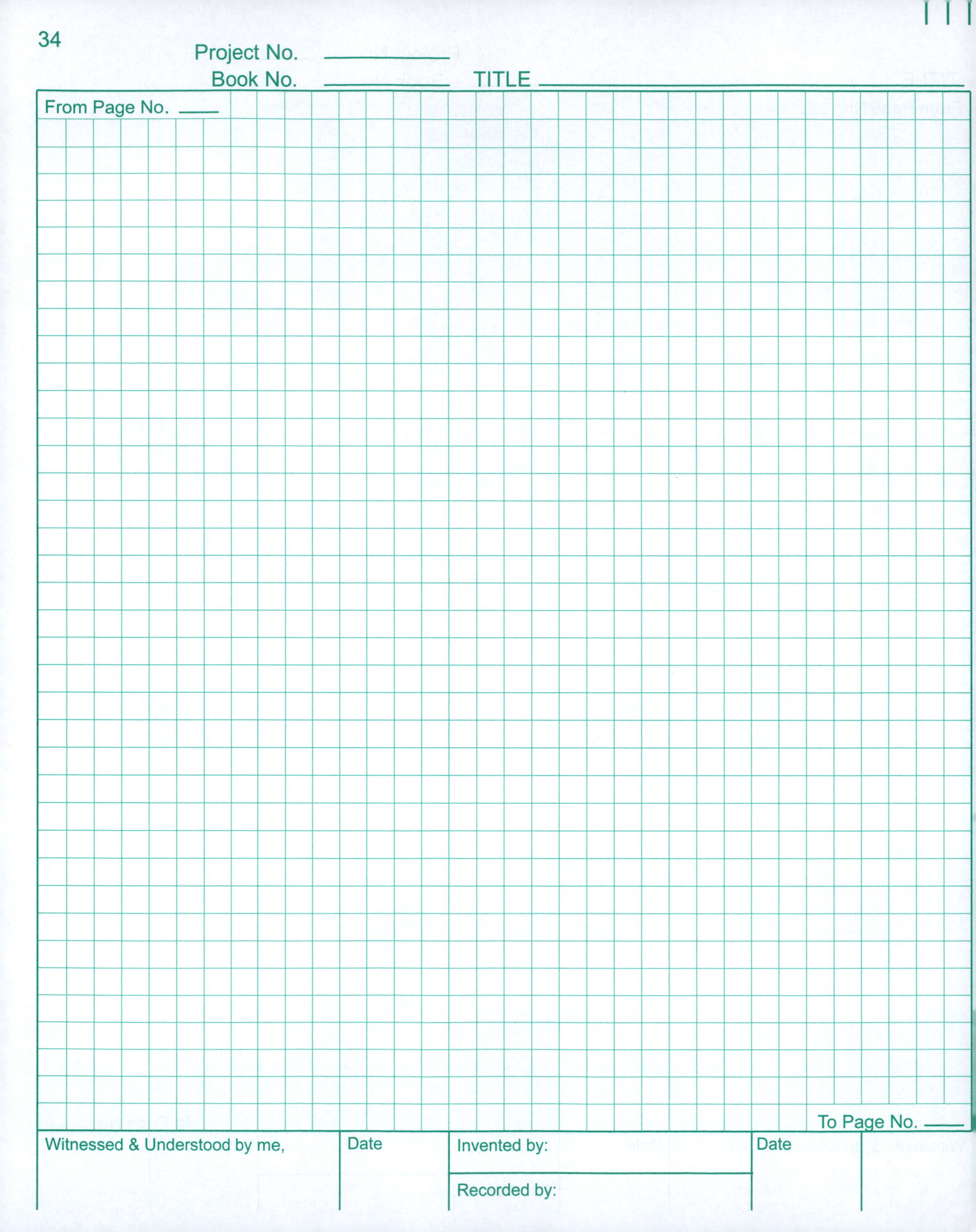
Project No.
Book No.
TITLE
From Page No.
To Page No.
Witnessed & Understood by me,
Date
Invented by:
Recorded by:
Date

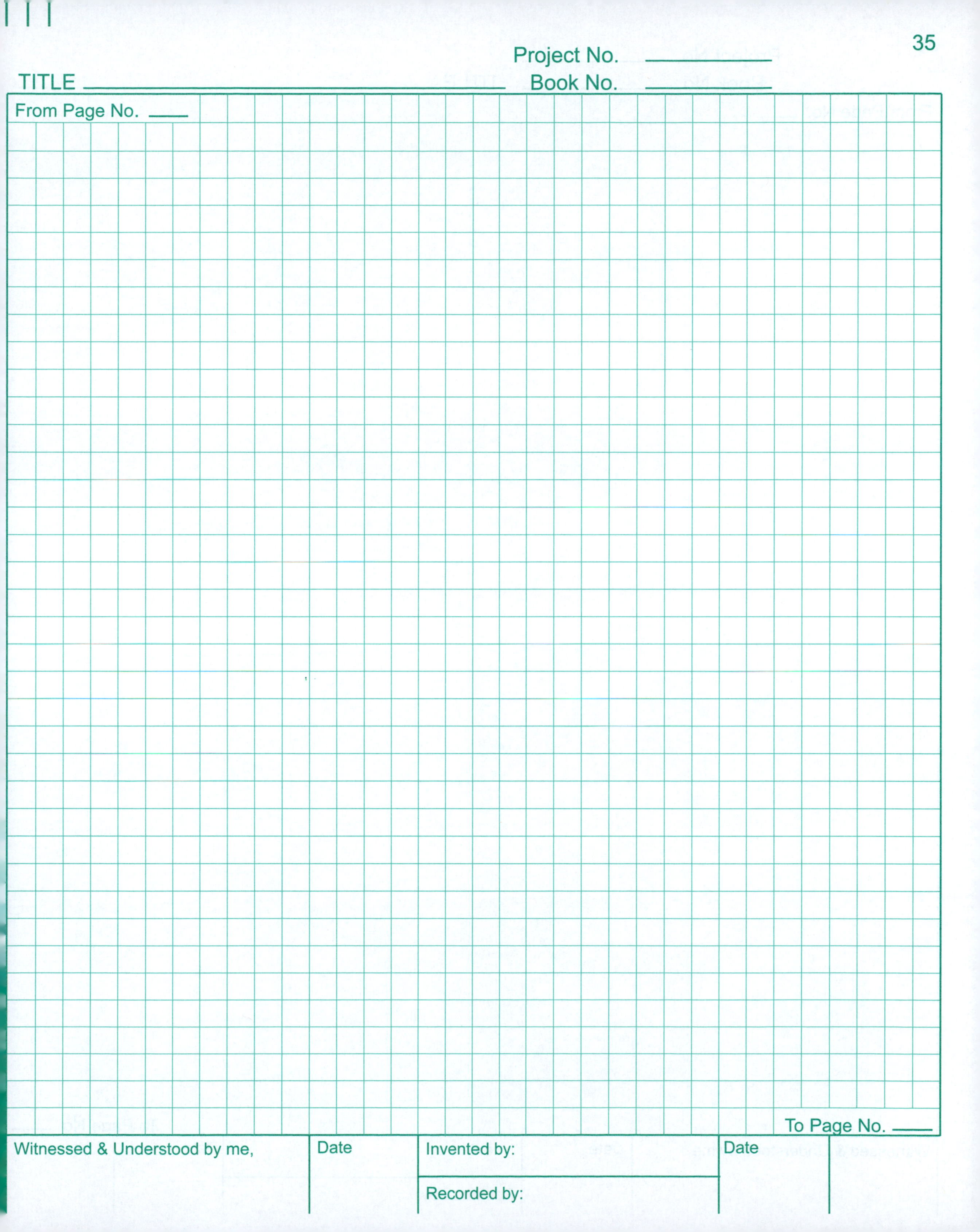

Project No. ____________

TITLE ____________ Book No. ____________

From Page No. ____

To Page No. ____

Witnessed & Understood by me,	Date	Invented by:	Date
		Recorded by:	

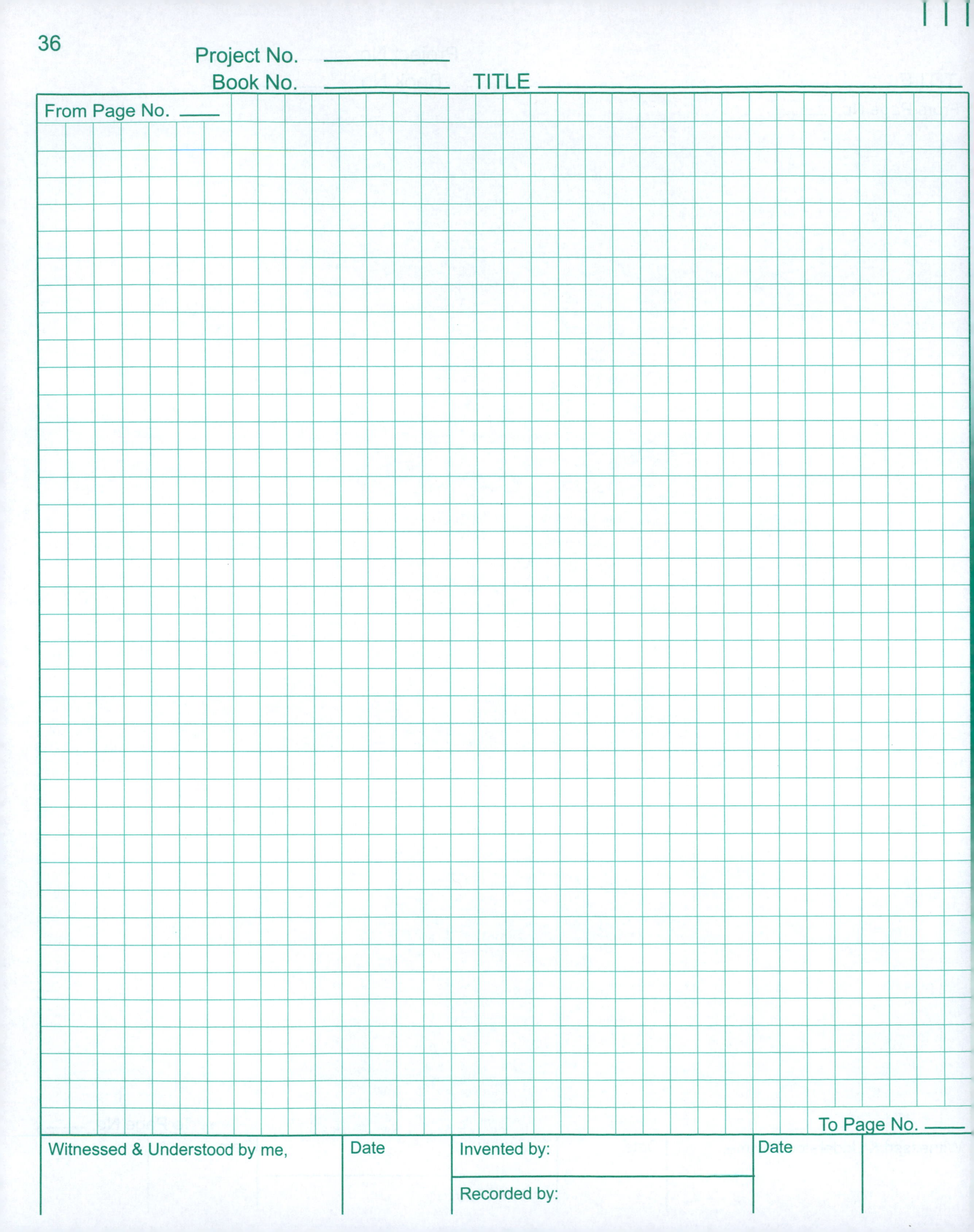
Project No.
Book No.
TITLE
From Page No.
To Page No.
Witnessed & Understood by me,
Date
Invented by:
Recorded by:
Date

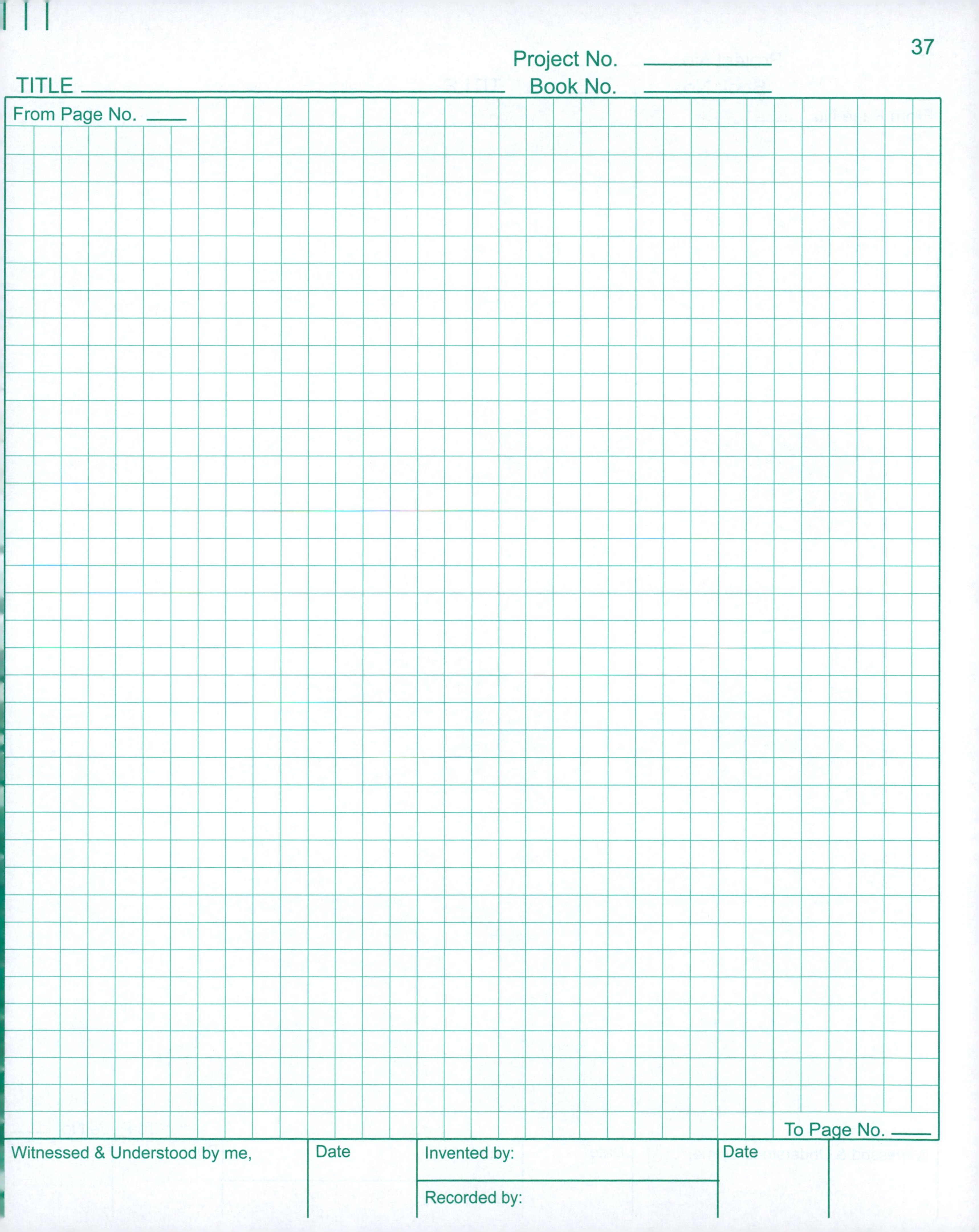
37
Project No.
TITLE
Book No.
From Page No.
To Page No.
Witnessed & Understood by me,
Date
Invented by:
Recorded by:
Date

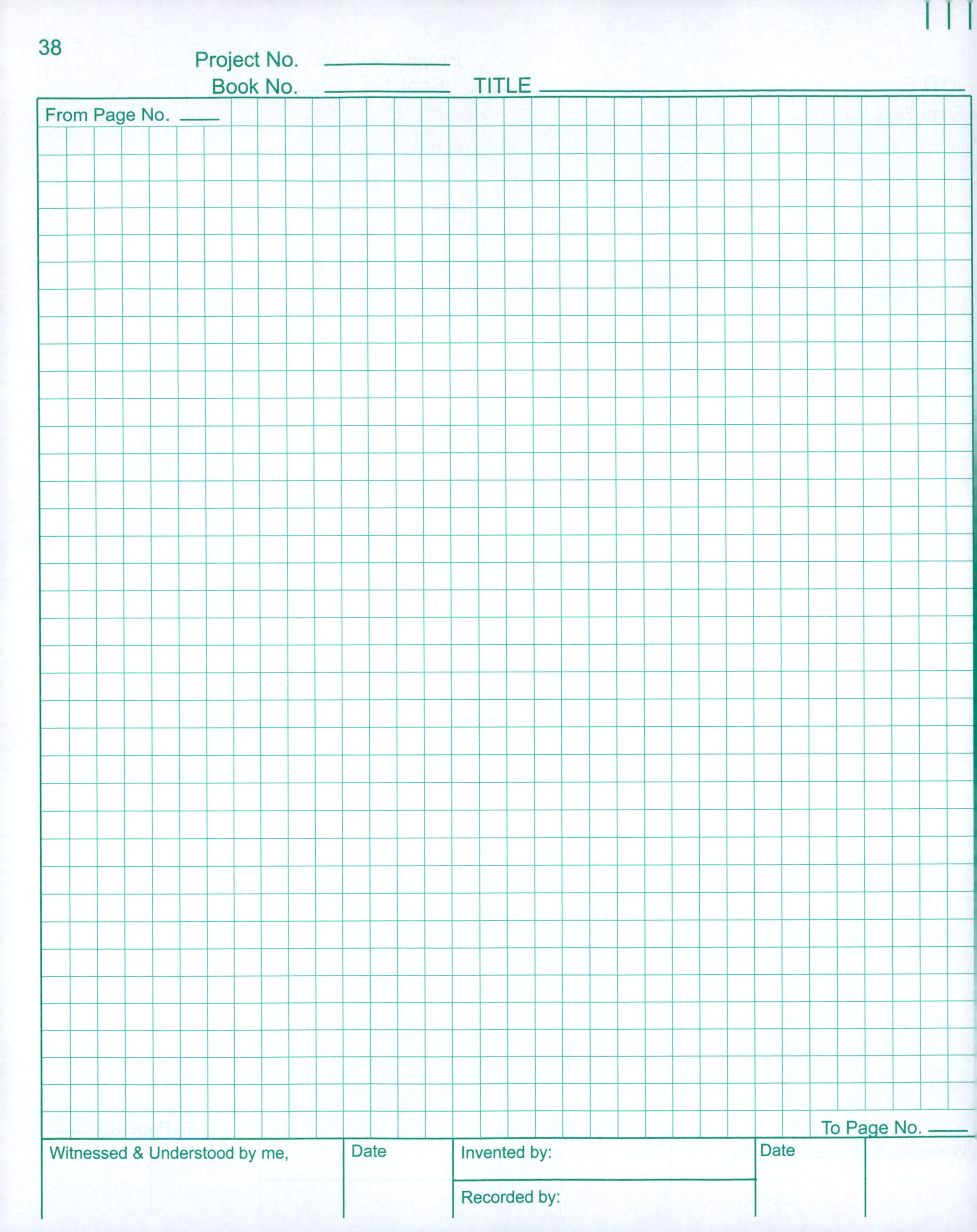

Project No. ____________

Book No. ____________ TITLE ____________

From Page No. ____

To Page No. ____

Witnessed & Understood by me,	Date	Invented by:	Date	
		Recorded by:		

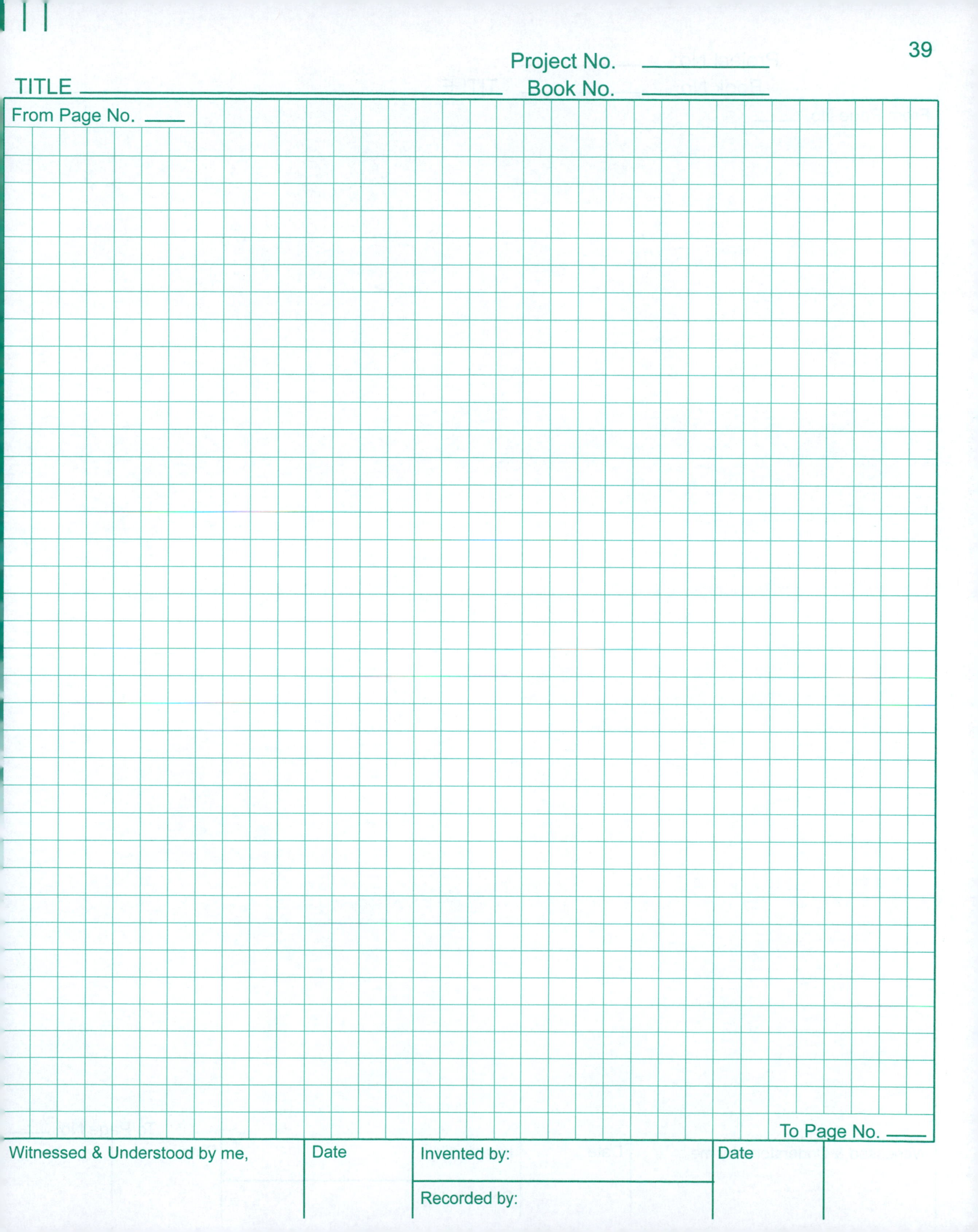

Project No. ____________

TITLE ________________________________ Book No. ____________

From Page No. ____

To Page No. ____

Witnessed & Understood by me,	Date	Invented by:	Date
		Recorded by:	

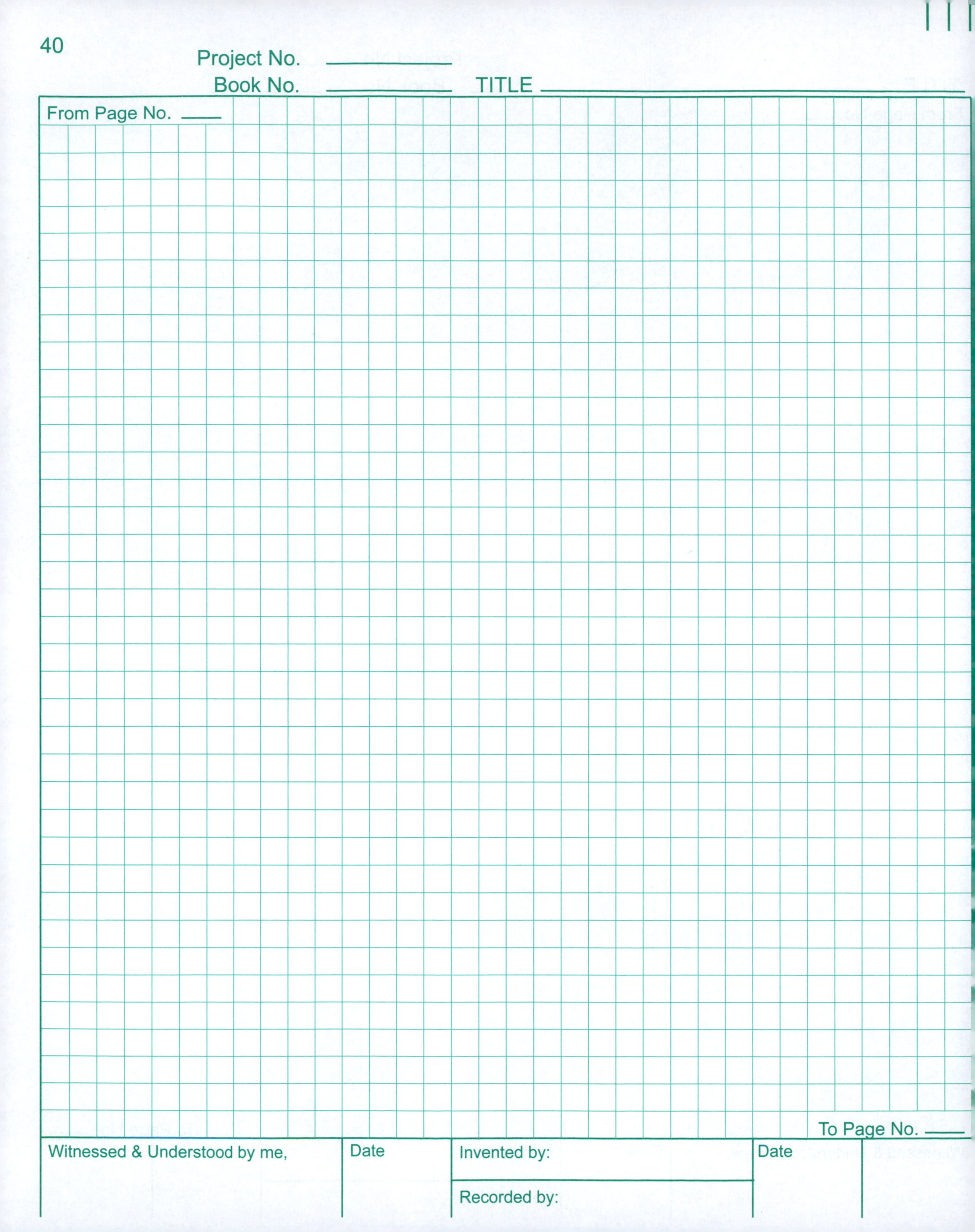
Project No.
Book No.
TITLE
From Page No.
To Page No.
Witnessed & Understood by me,
Date
Invented by:
Recorded by:
Date

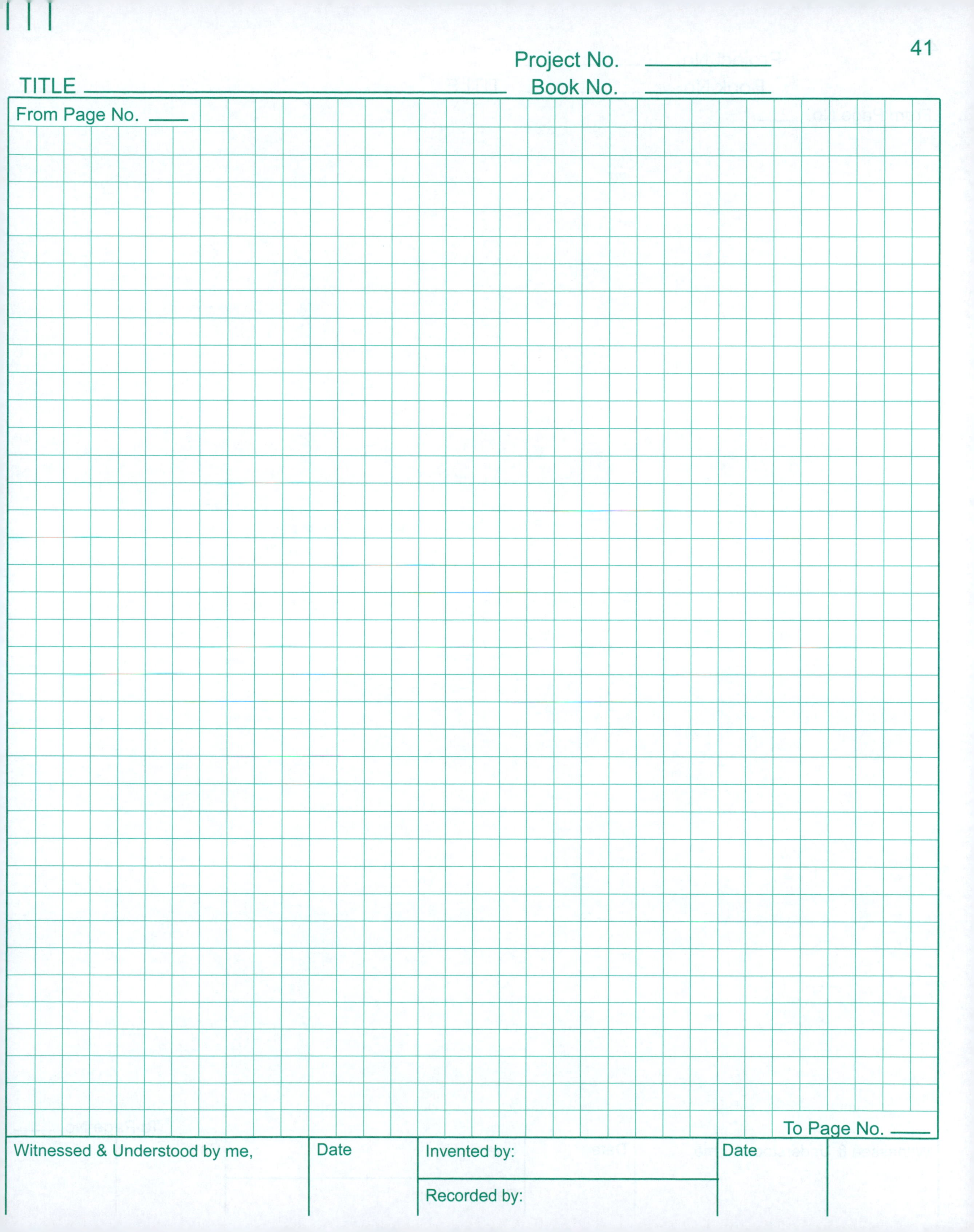
Project No.
TITLE
Book No.
From Page No.
To Page No.
Witnessed & Understood by me,
Date
Invented by:
Recorded by:
Date

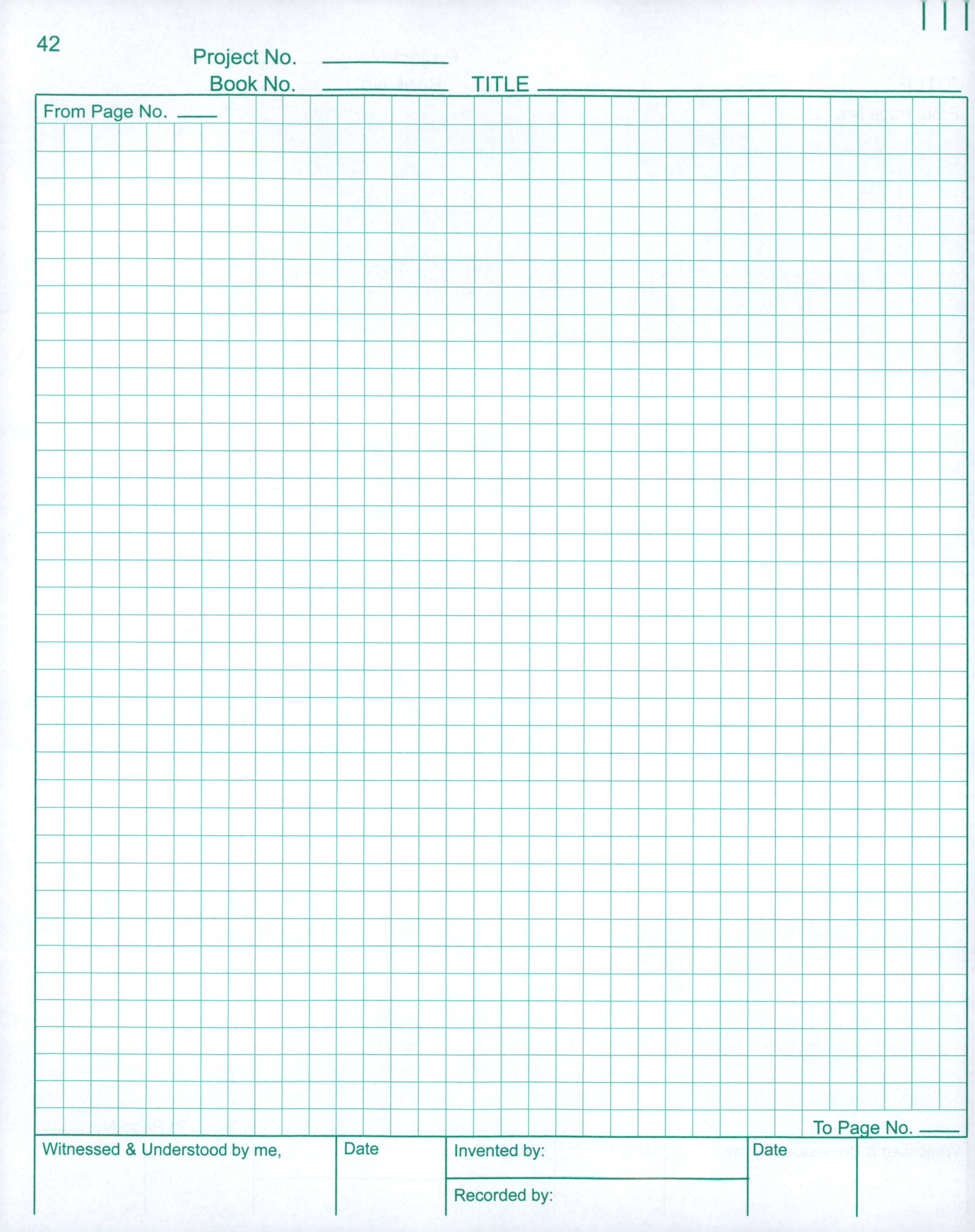
Project No.
Book No.
TITLE
From Page No.
To Page No.
Witnessed & Understood by me,
Date
Invented by:
Recorded by:
Date

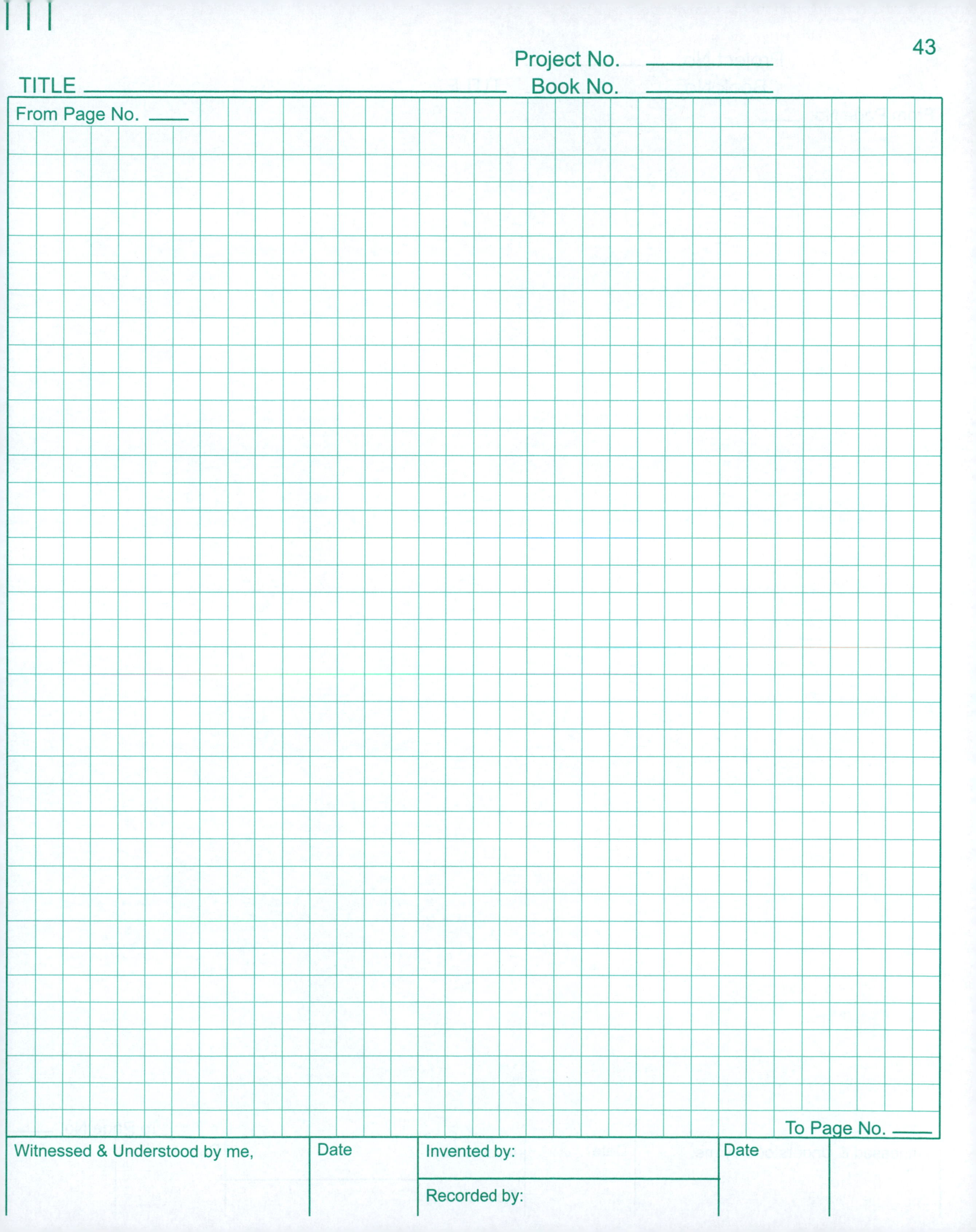
Project No.
TITLE
Book No.
From Page No.
To Page No.
Witnessed & Understood by me,
Date
Invented by:
Recorded by:
Date

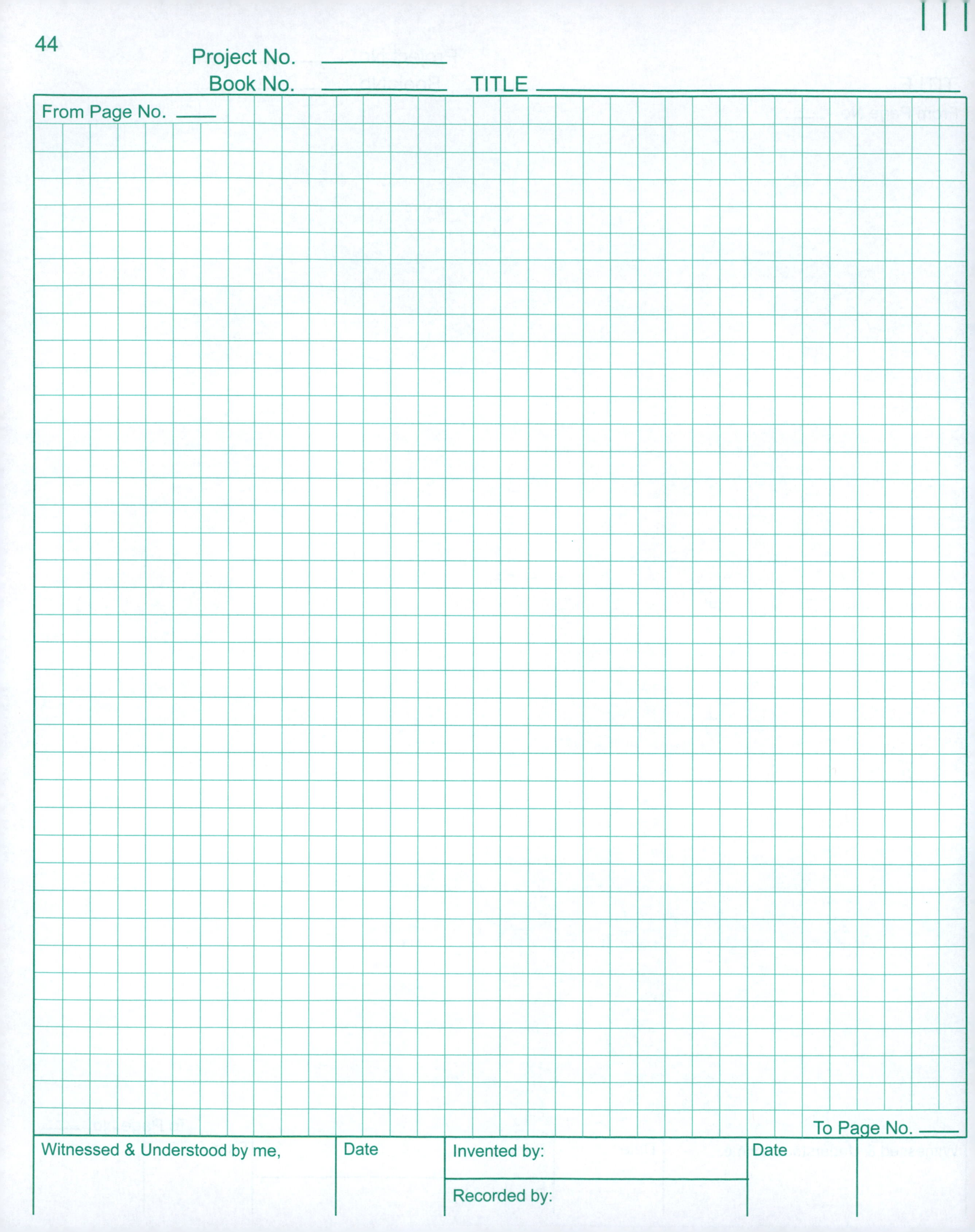

Project No. ____________

Book No. ____________ TITLE ____________

From Page No. ____

To Page No. ____

Witnessed & Understood by me,

Date

Invented by:

Recorded by:

Date

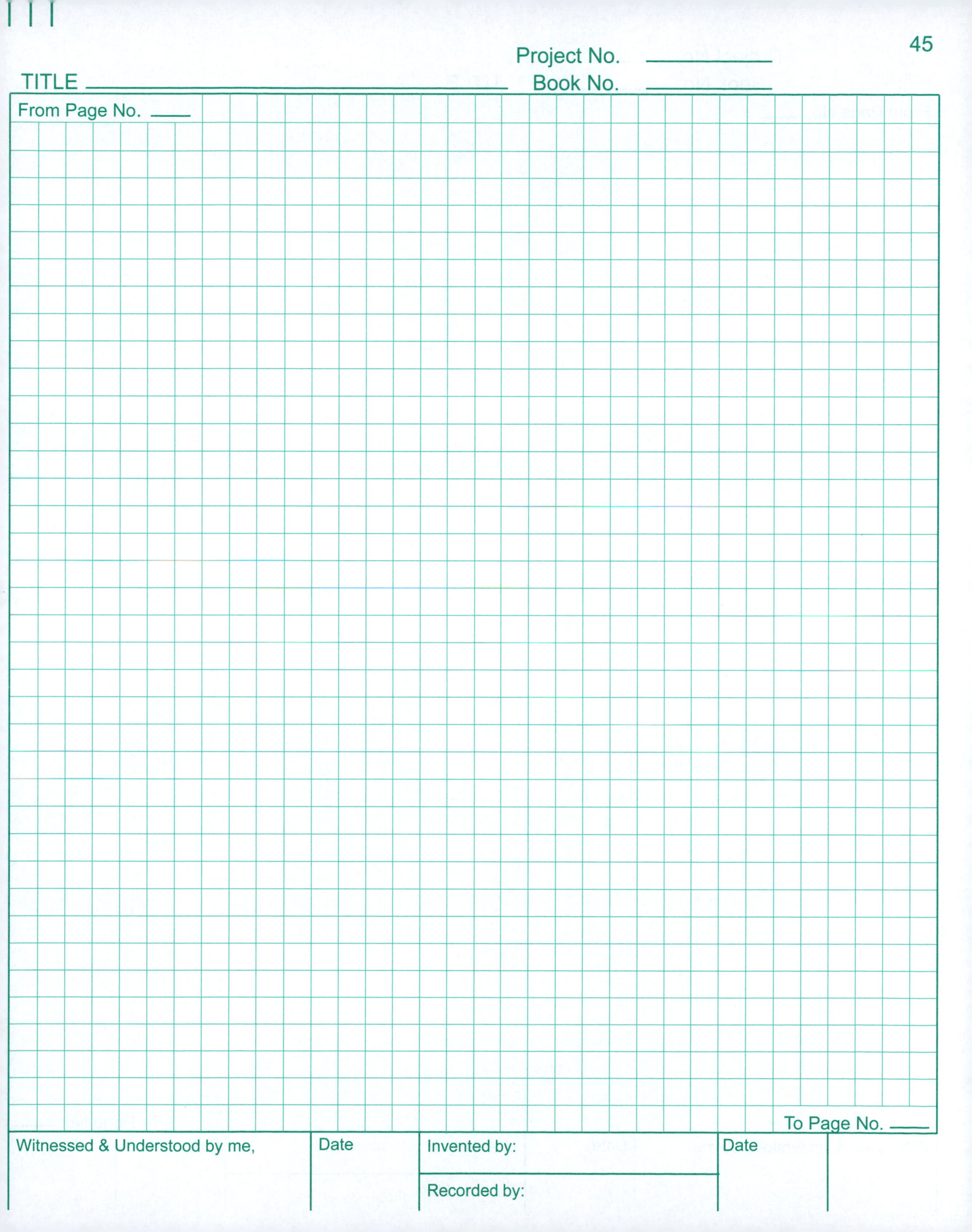
Project No.
TITLE
Book No.
From Page No.
To Page No.
Witnessed & Understood by me,
Date
Invented by:
Recorded by:
Date

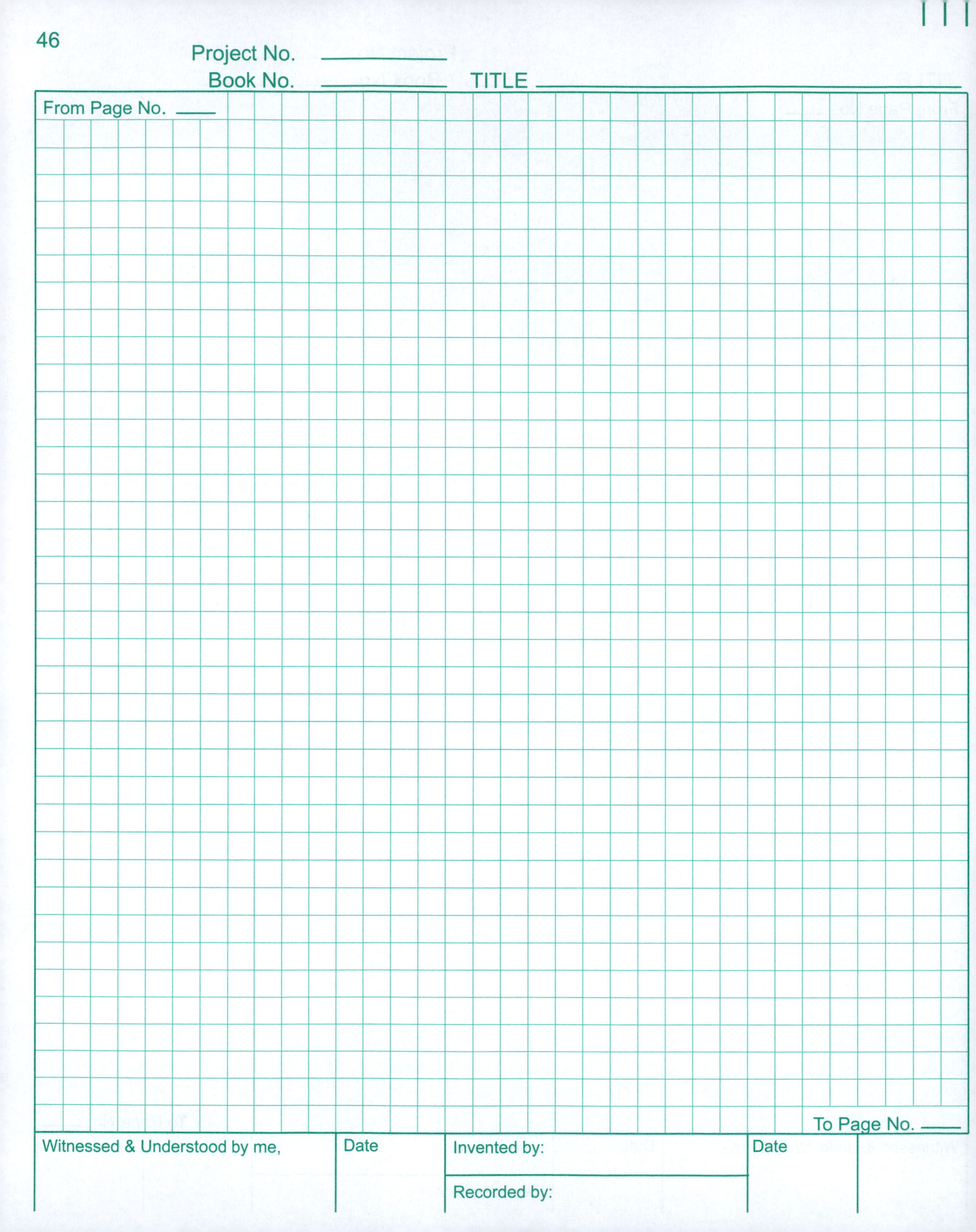
Project No.
Book No.
TITLE
From Page No.
To Page No.
Witnessed & Understood by me,
Date
Invented by:
Recorded by:
Date

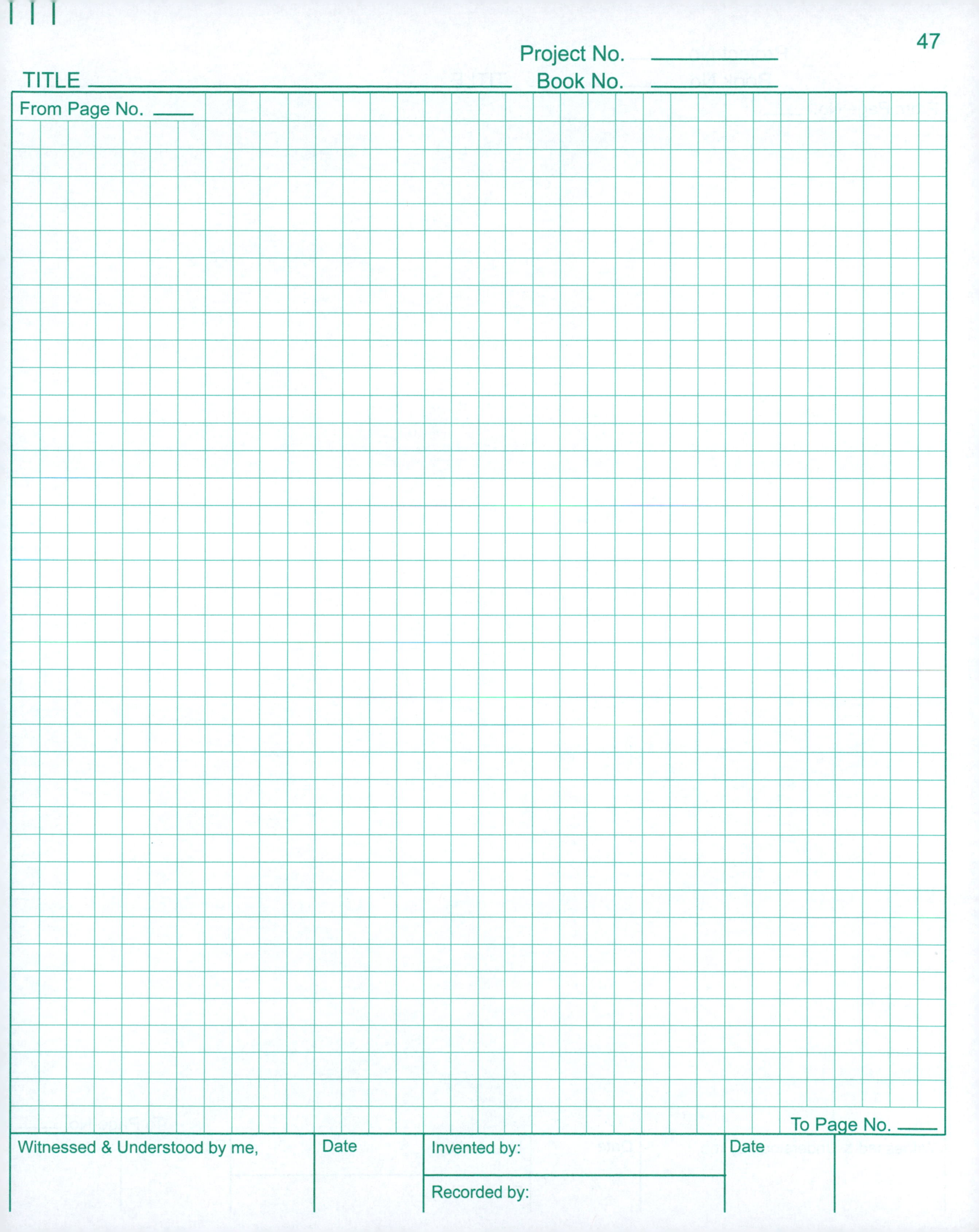
Project No.
TITLE
Book No.
From Page No.
To Page No.
Witnessed & Understood by me,
Date
Invented by:
Recorded by:
Date

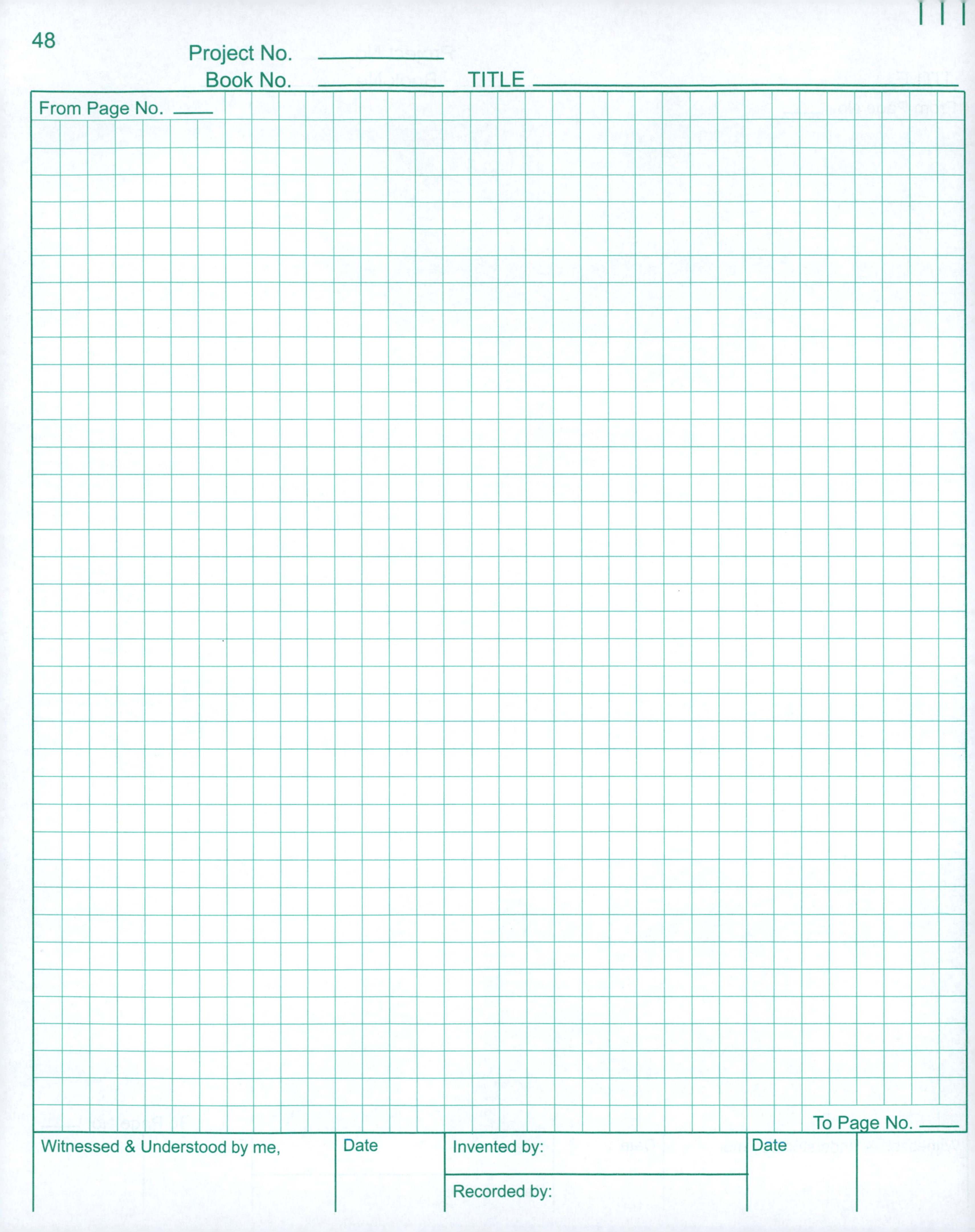
Project No.
Book No.
TITLE
From Page No.
To Page No.
Witnessed & Understood by me,
Date
Invented by:
Recorded by:
Date

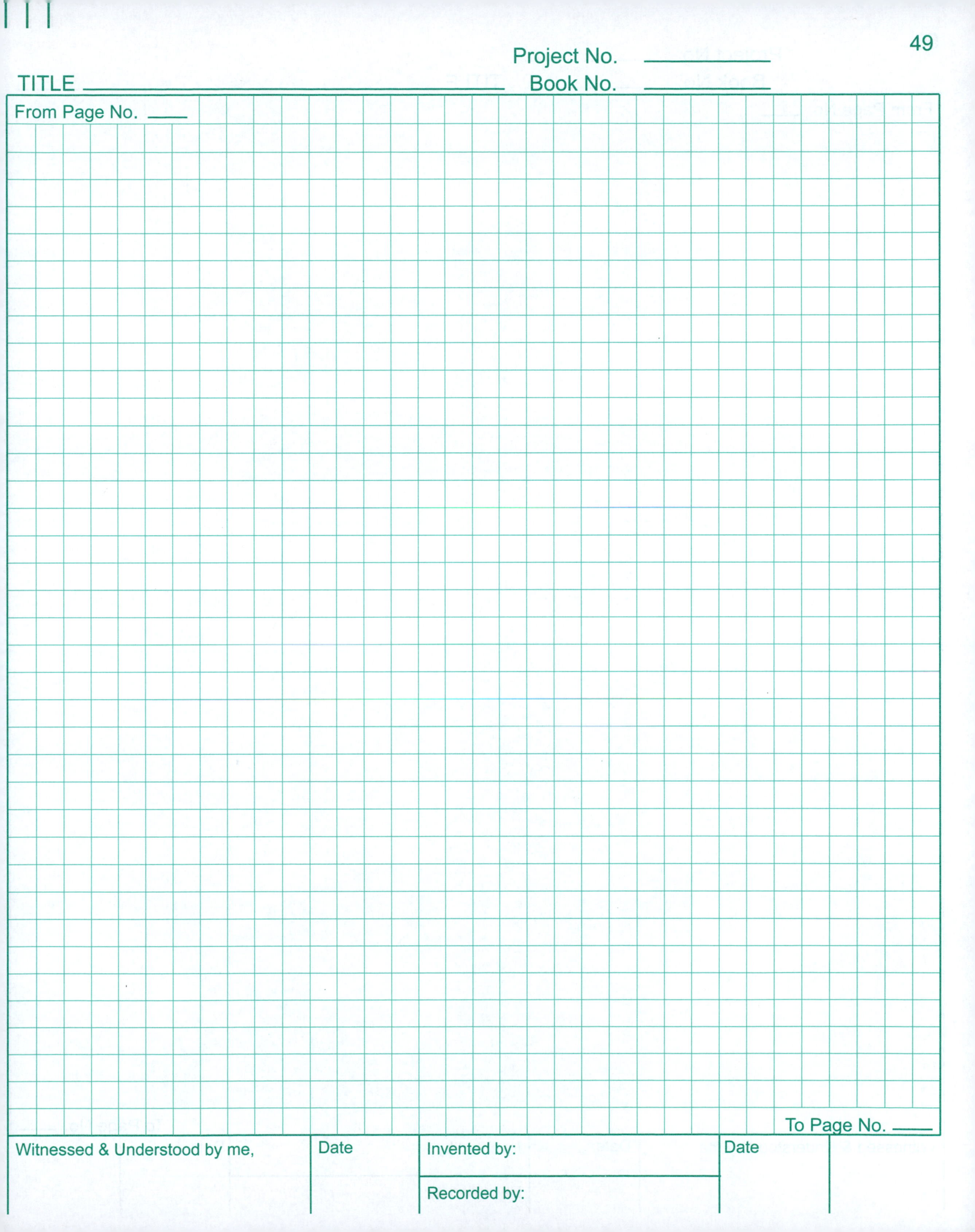
Project No.
TITLE
Book No.
From Page No.
To Page No.
Witnessed & Understood by me,
Date
Invented by:
Recorded by:
Date

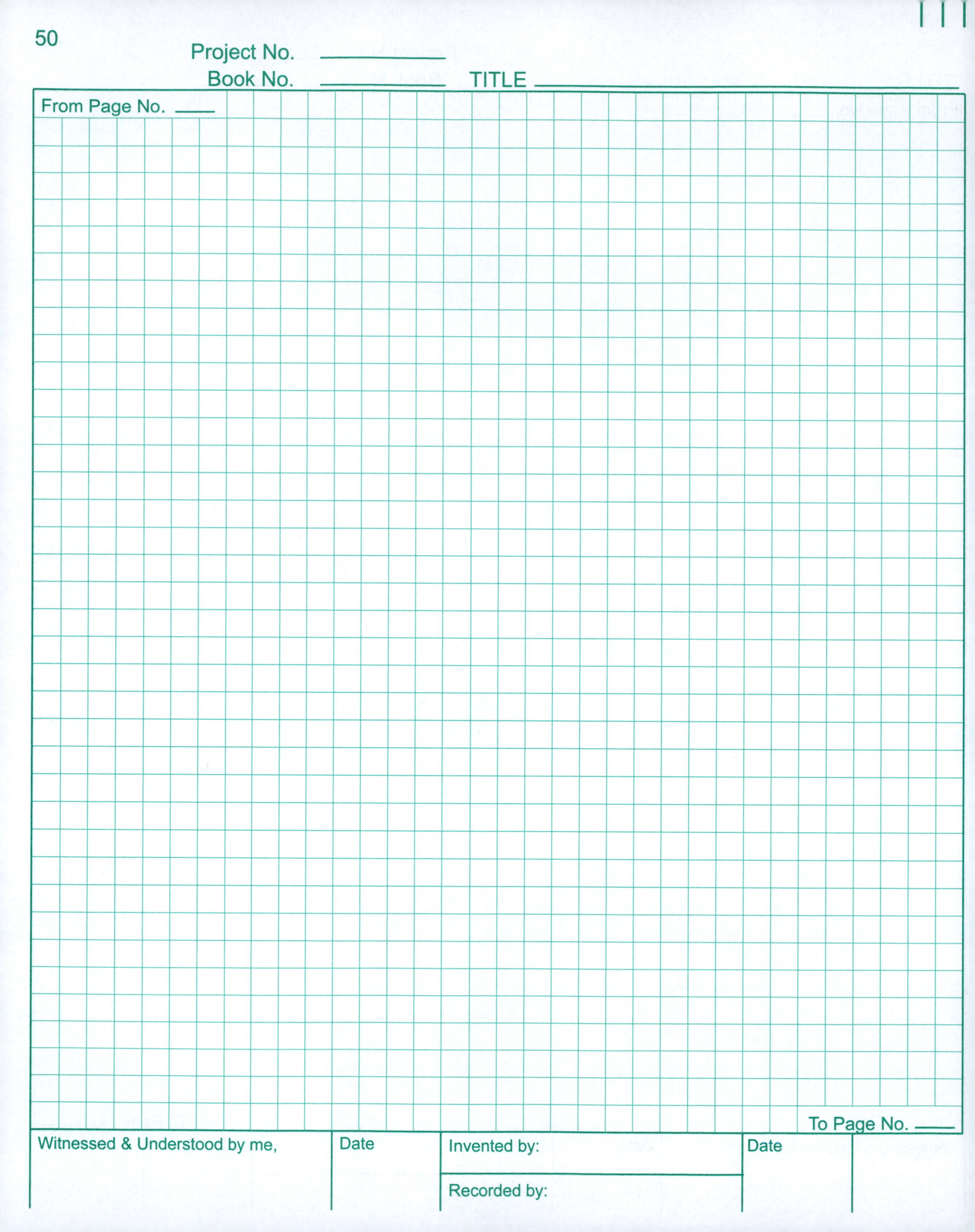

Project No. ______

Book No. ______ TITLE ______

From Page No. ____

To Page No. ____

Witnessed & Understood by me,	Date	Invented by:	Date
		Recorded by:	

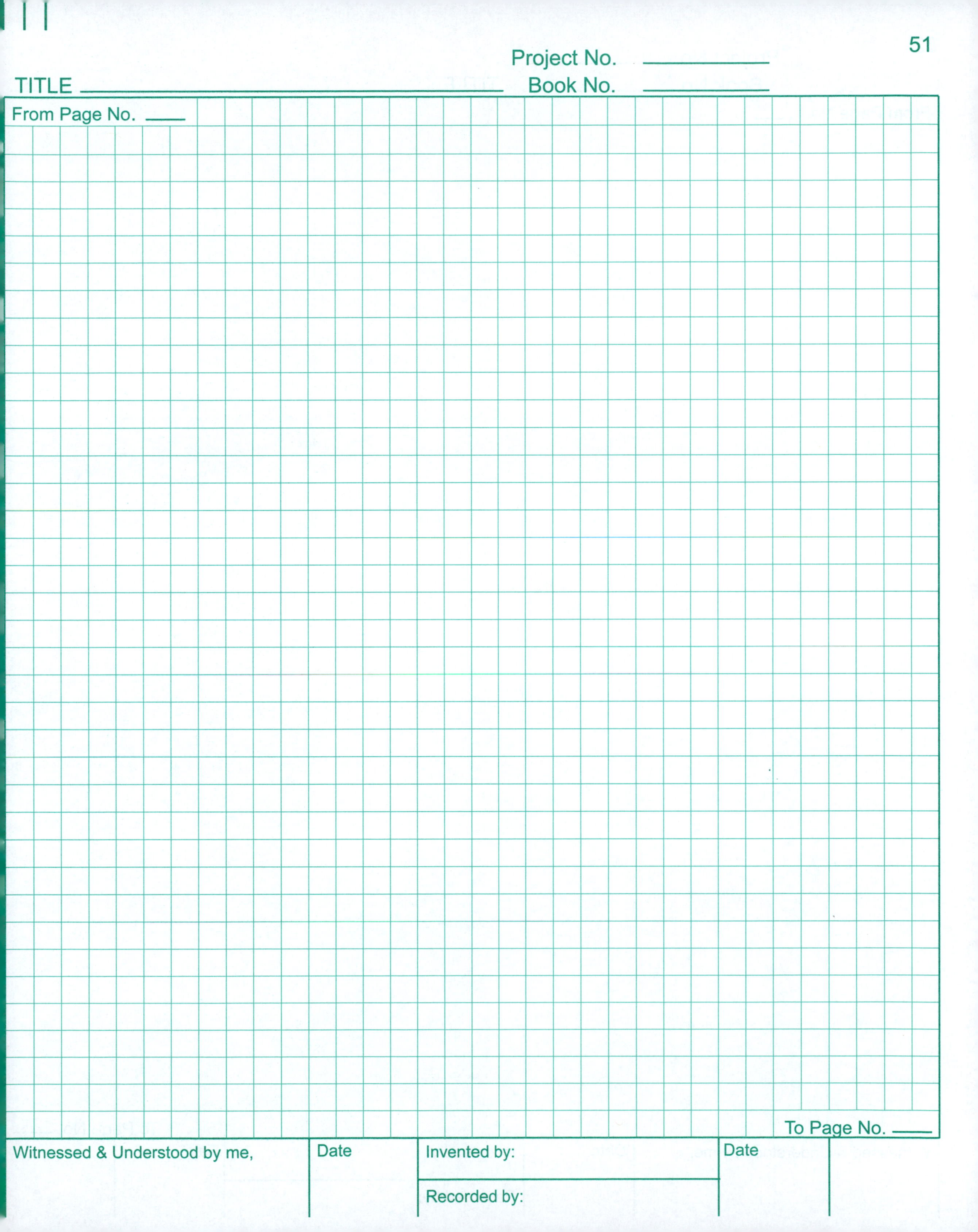
Project No.
TITLE
Book No.
From Page No.
To Page No.
Witnessed & Understood by me,
Date
Invented by:
Recorded by:
Date

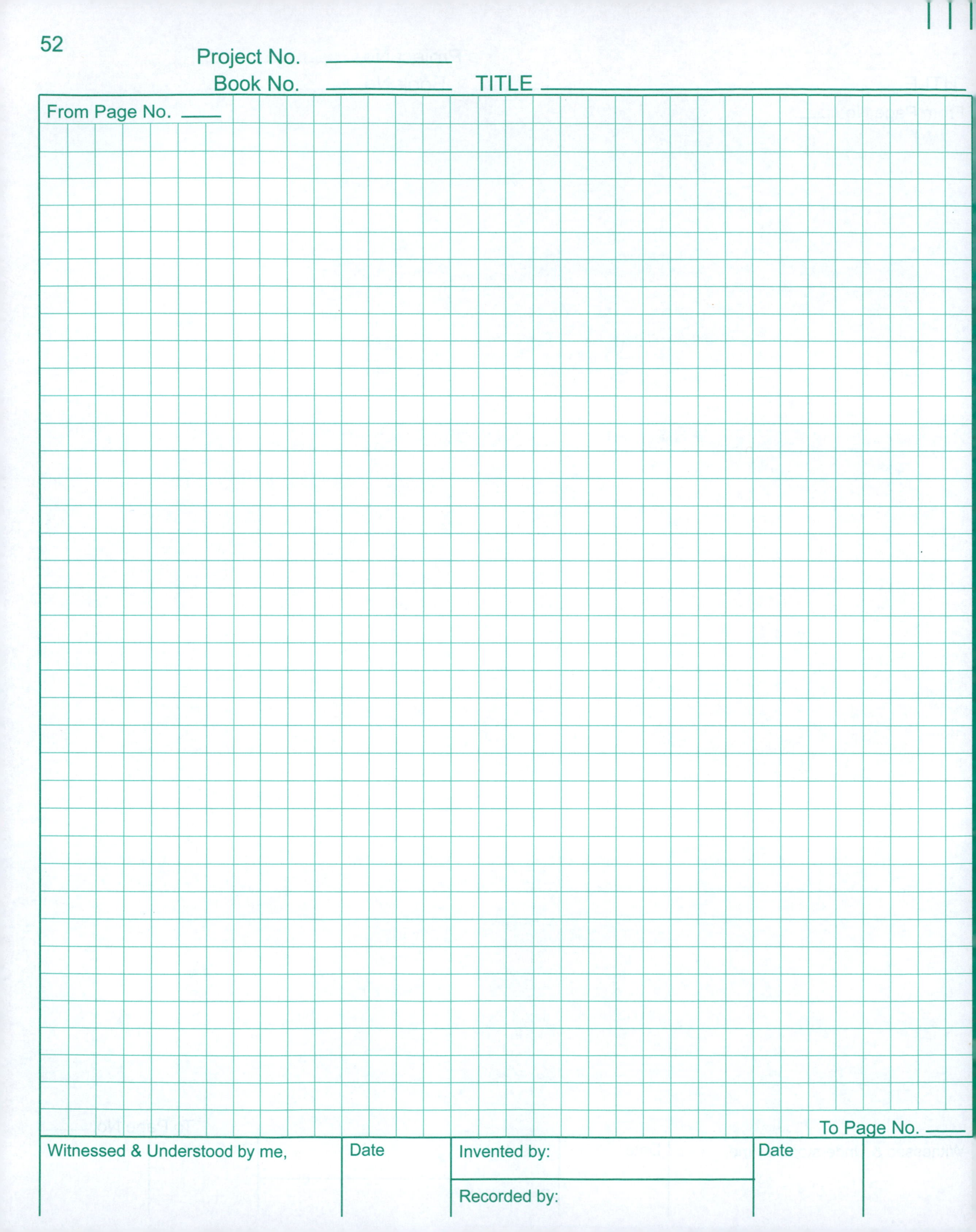
Project No.
Book No.
TITLE
From Page No.
To Page No.
Witnessed & Understood by me,
Date
Invented by:
Recorded by:
Date

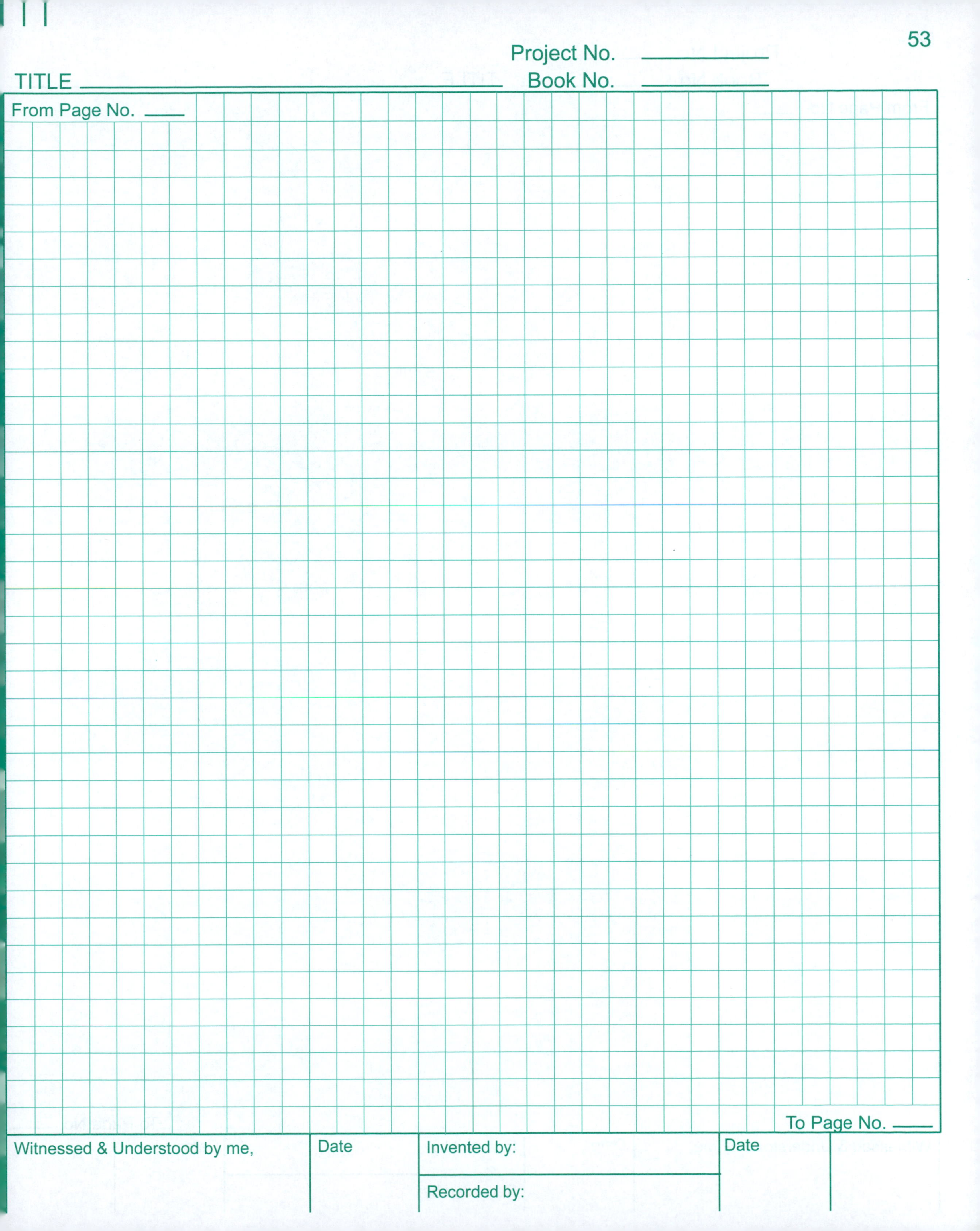
Project No.
TITLE
Book No.
From Page No.
To Page No.
Witnessed & Understood by me,
Date
Invented by:
Recorded by:
Date

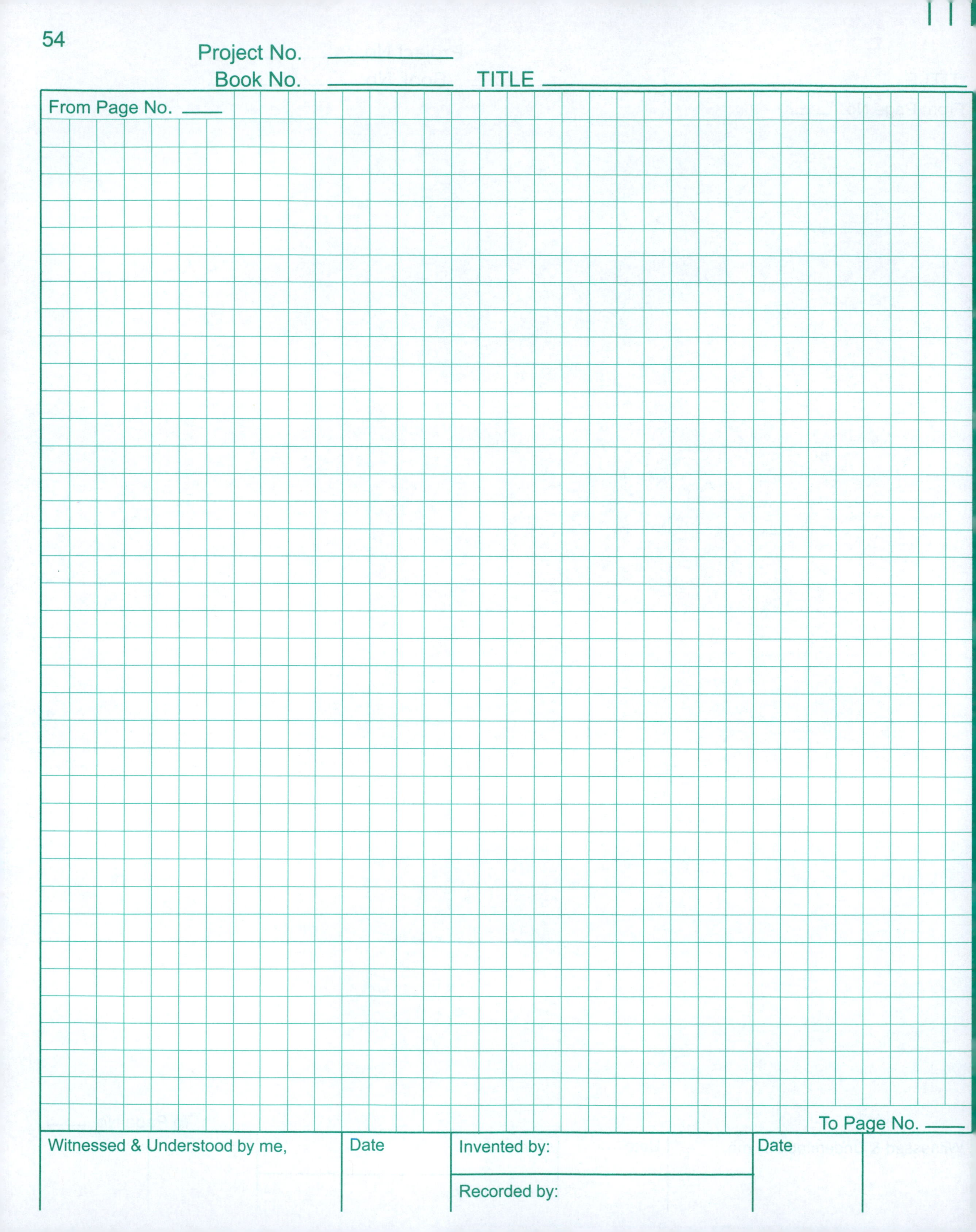
54
Project No.
Book No.
TITLE
From Page No.
To Page No.
Witnessed & Understood by me,
Date
Invented by:
Recorded by:
Date

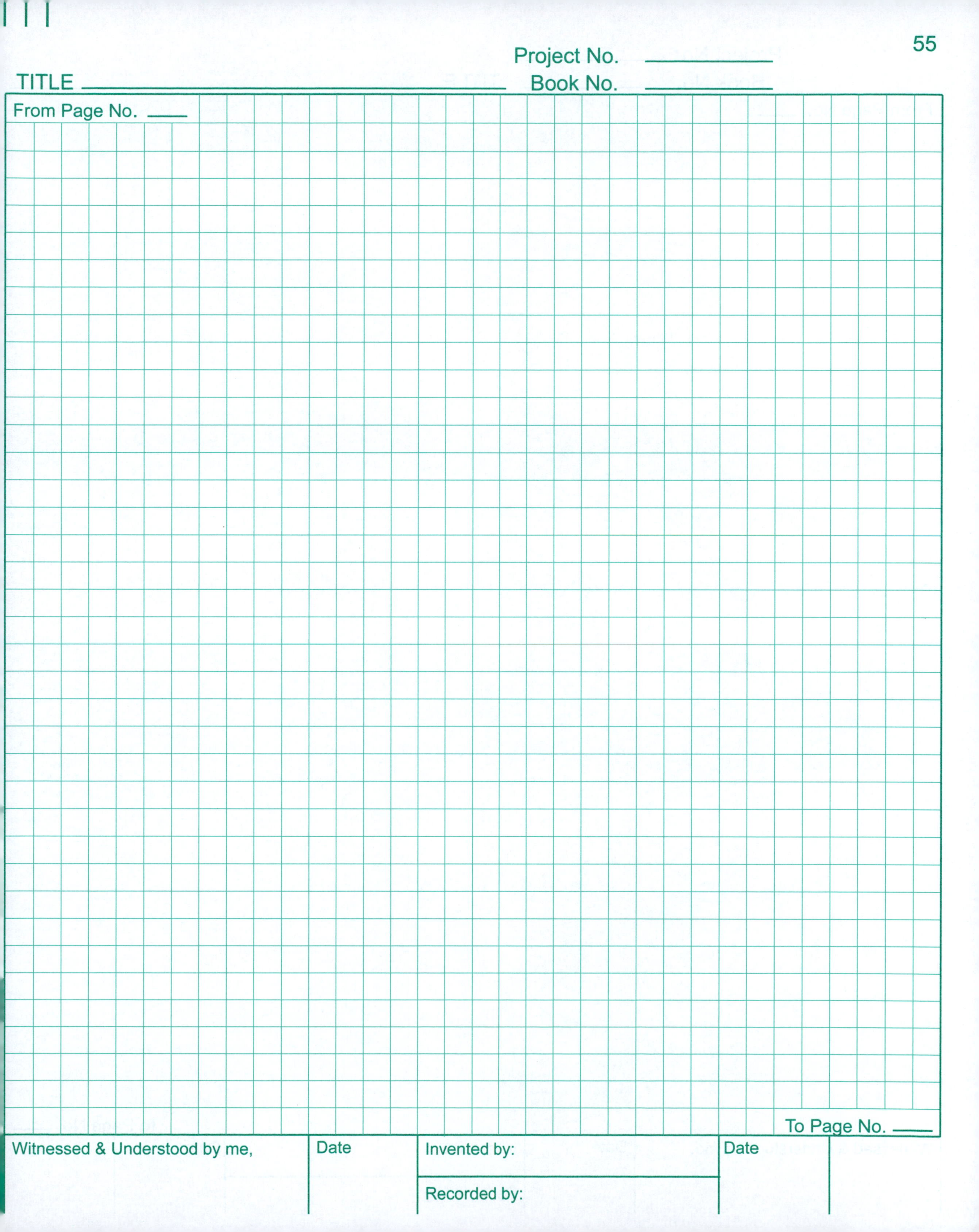

Project No. ____________

TITLE ________________________________ Book No. ____________

From Page No. ____

To Page No. ____

Witnessed & Understood by me,	Date	Invented by:	Date
		Recorded by:	

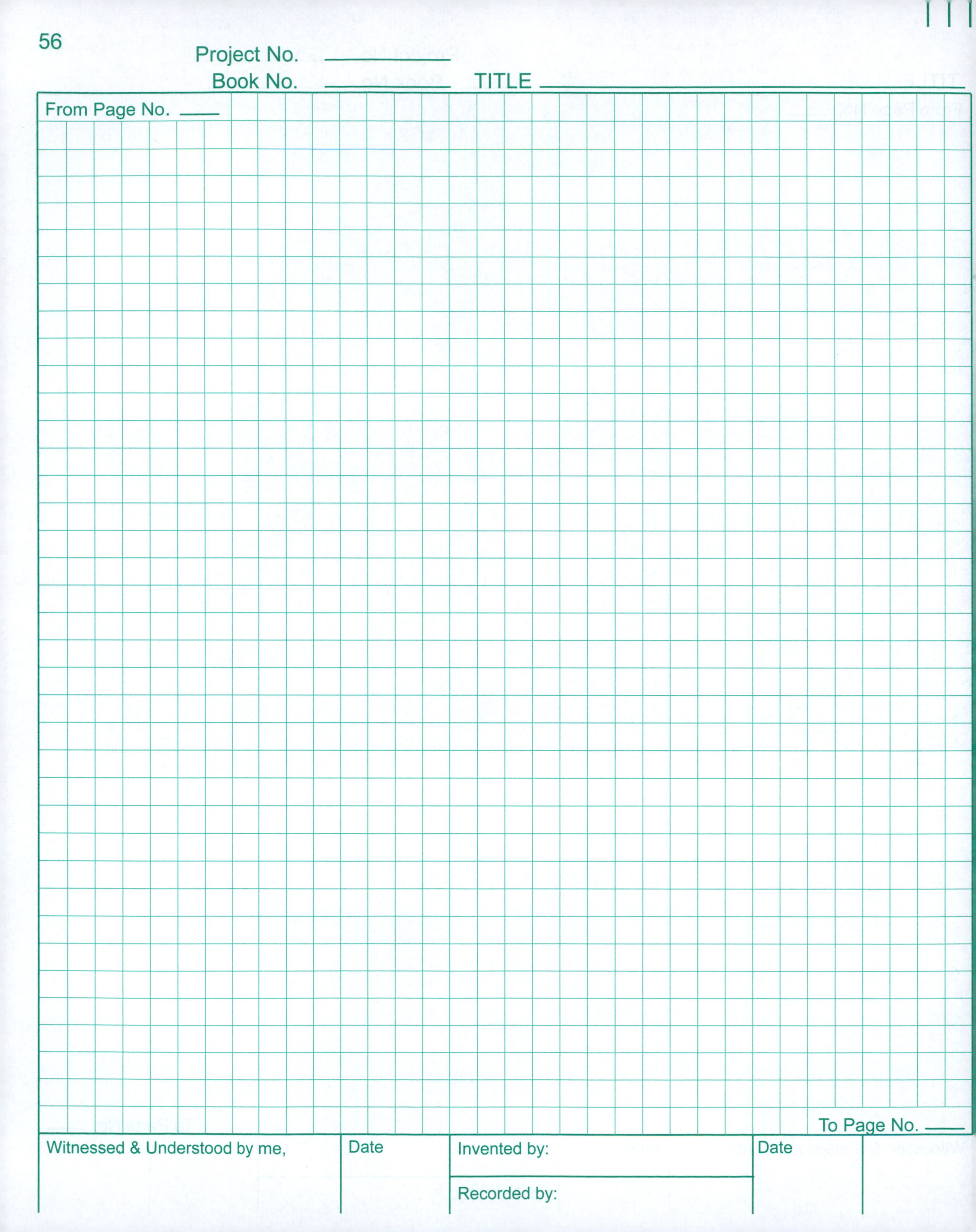
Project No.
Book No.
TITLE
From Page No.
To Page No.
Witnessed & Understood by me,
Date
Invented by:
Recorded by:
Date

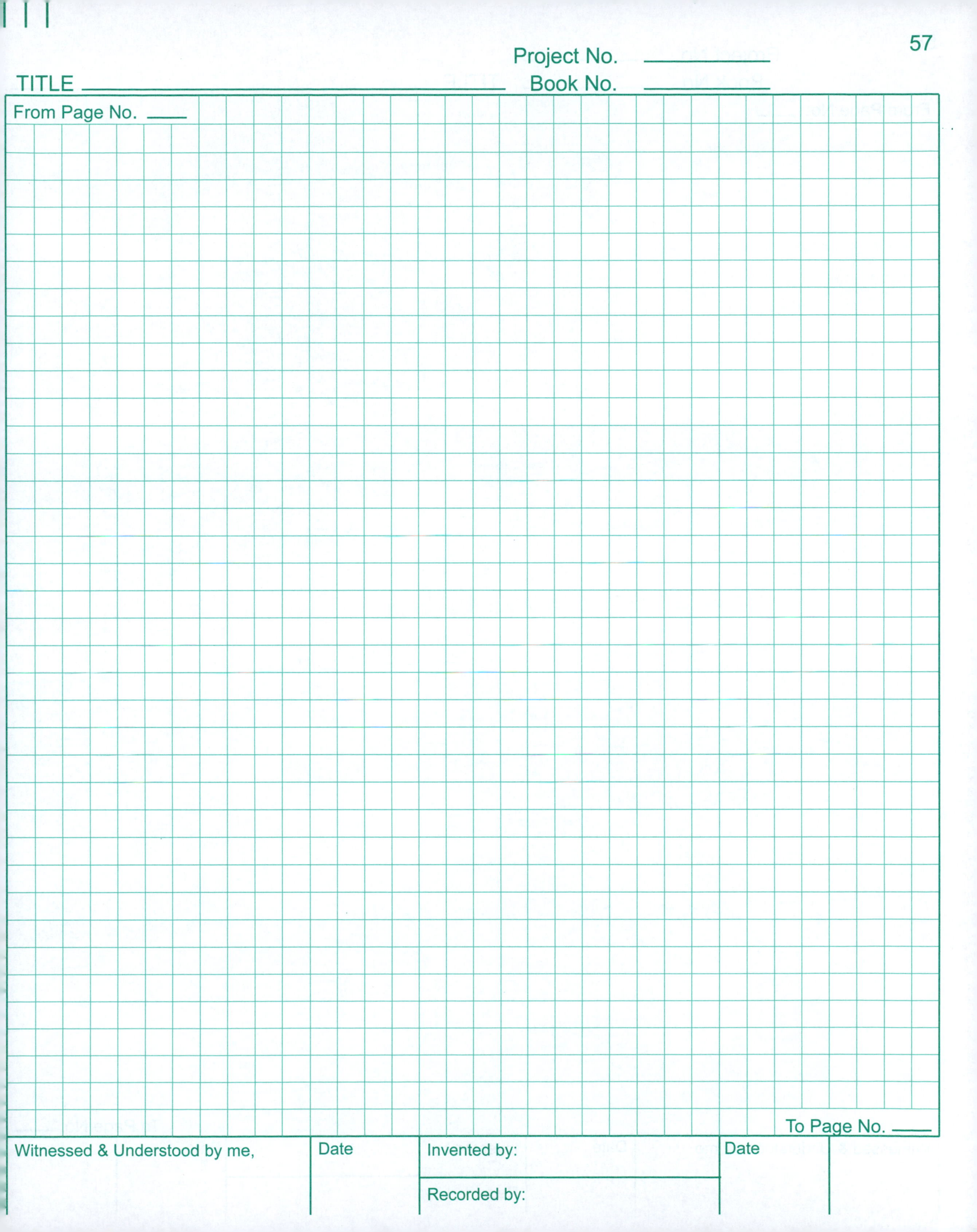
57
Project No.
TITLE
Book No.
From Page No.
To Page No.
Witnessed & Understood by me,
Date
Invented by:
Recorded by:
Date

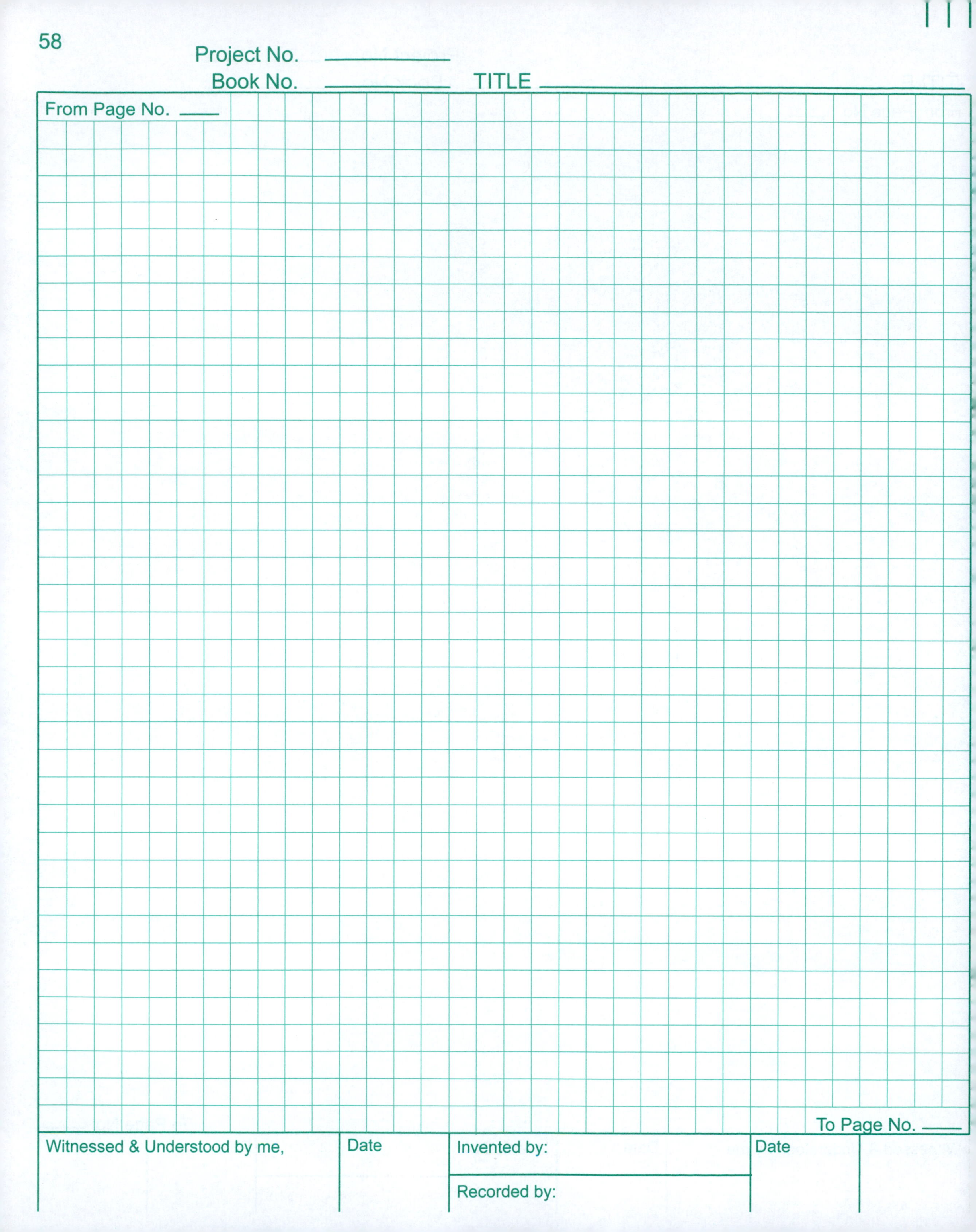
Project No.
Book No.
TITLE
From Page No.
To Page No.
Witnessed & Understood by me,
Date
Invented by:
Recorded by:
Date

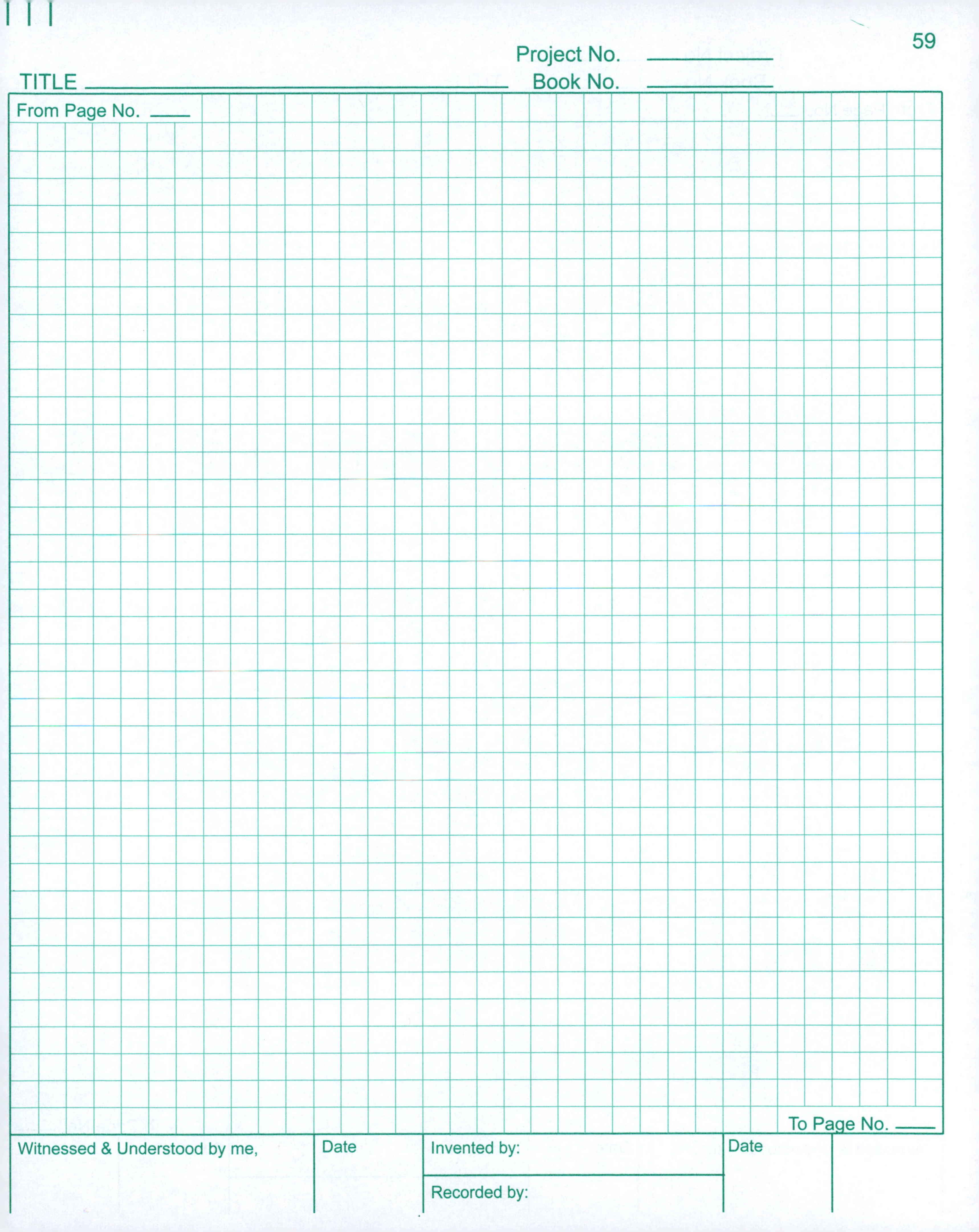
Project No.
TITLE
Book No.
From Page No.
To Page No.
Witnessed & Understood by me,
Date
Invented by:
Recorded by:
Date

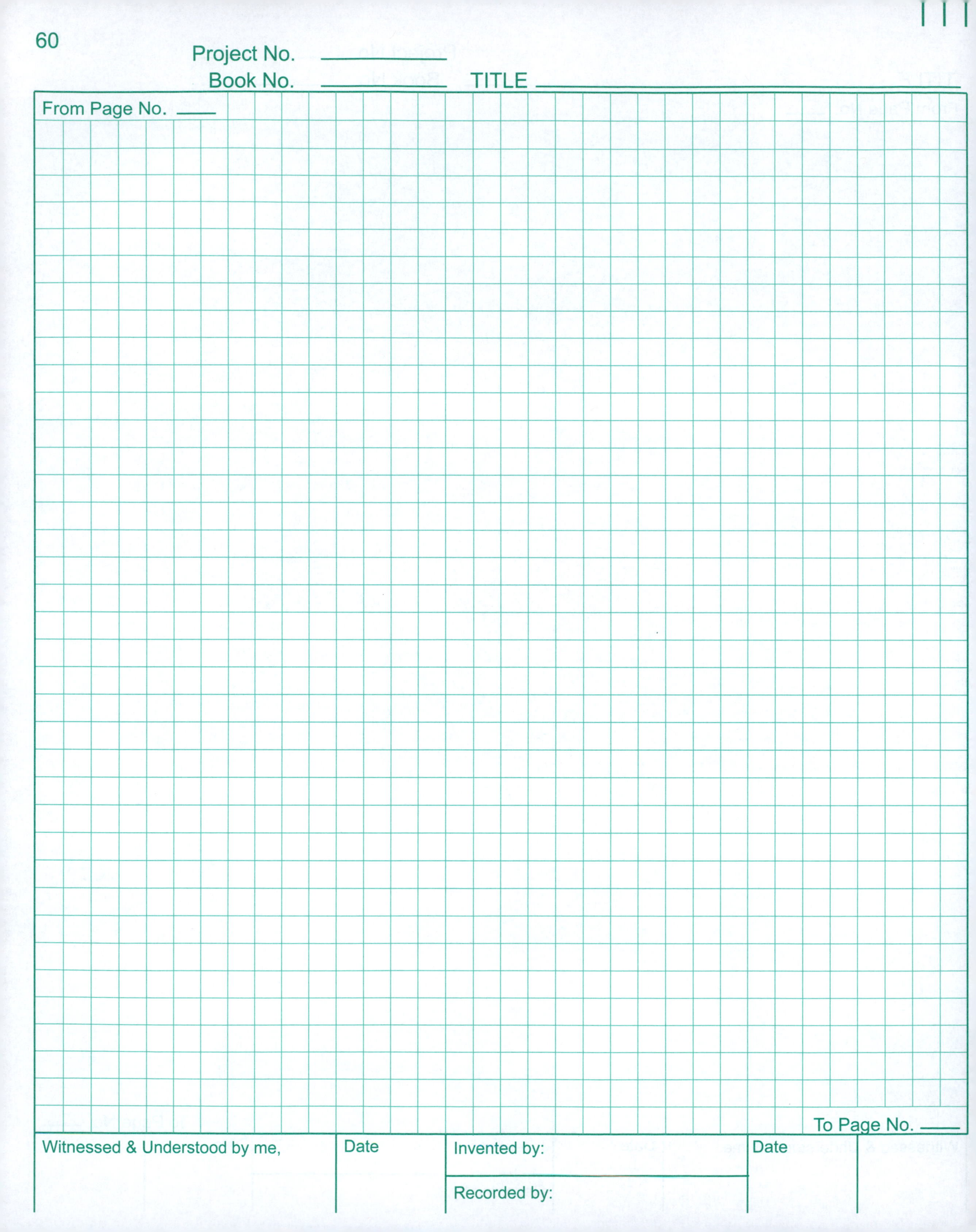
Project No.
Book No.
TITLE
From Page No.
To Page No.
Witnessed & Understood by me,
Date
Invented by:
Recorded by:
Date

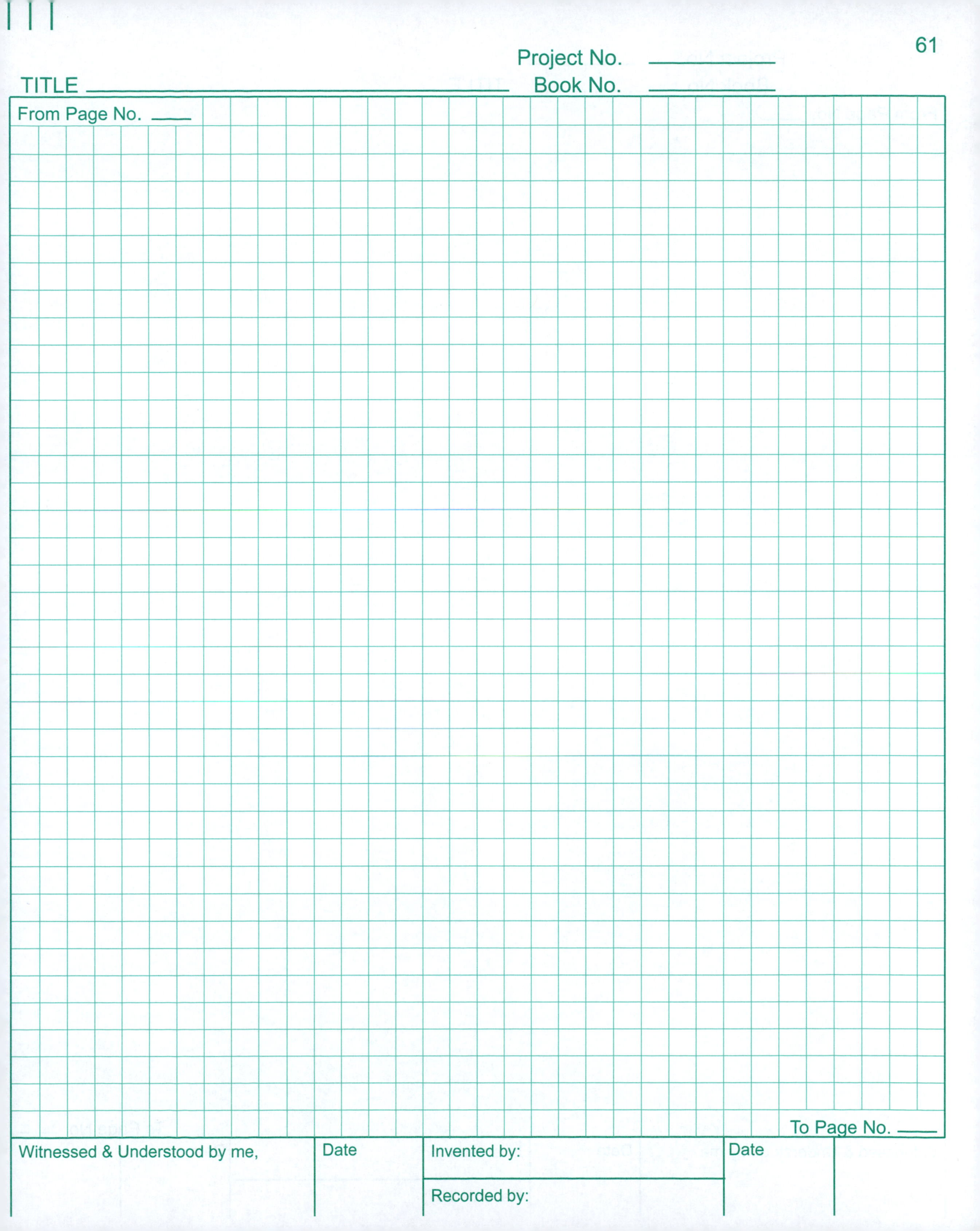

Project No. ____________

TITLE ________________________ Book No. ____________

From Page No. ____

To Page No. ____

Witnessed & Understood by me,	Date	Invented by:	Date
		Recorded by:	

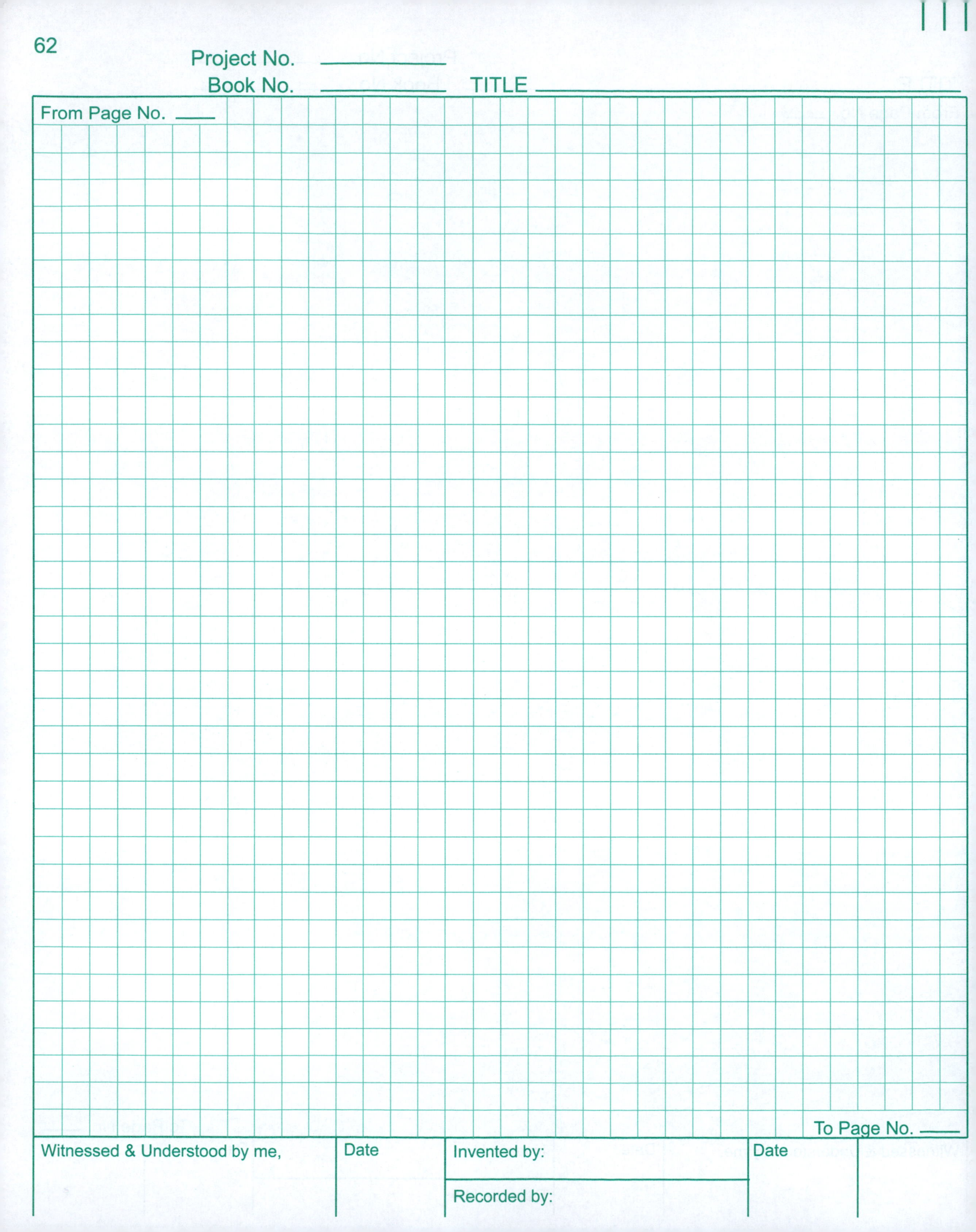
Project No.
Book No.
TITLE
From Page No.
To Page No.
Witnessed & Understood by me,
Date
Invented by:
Recorded by:
Date

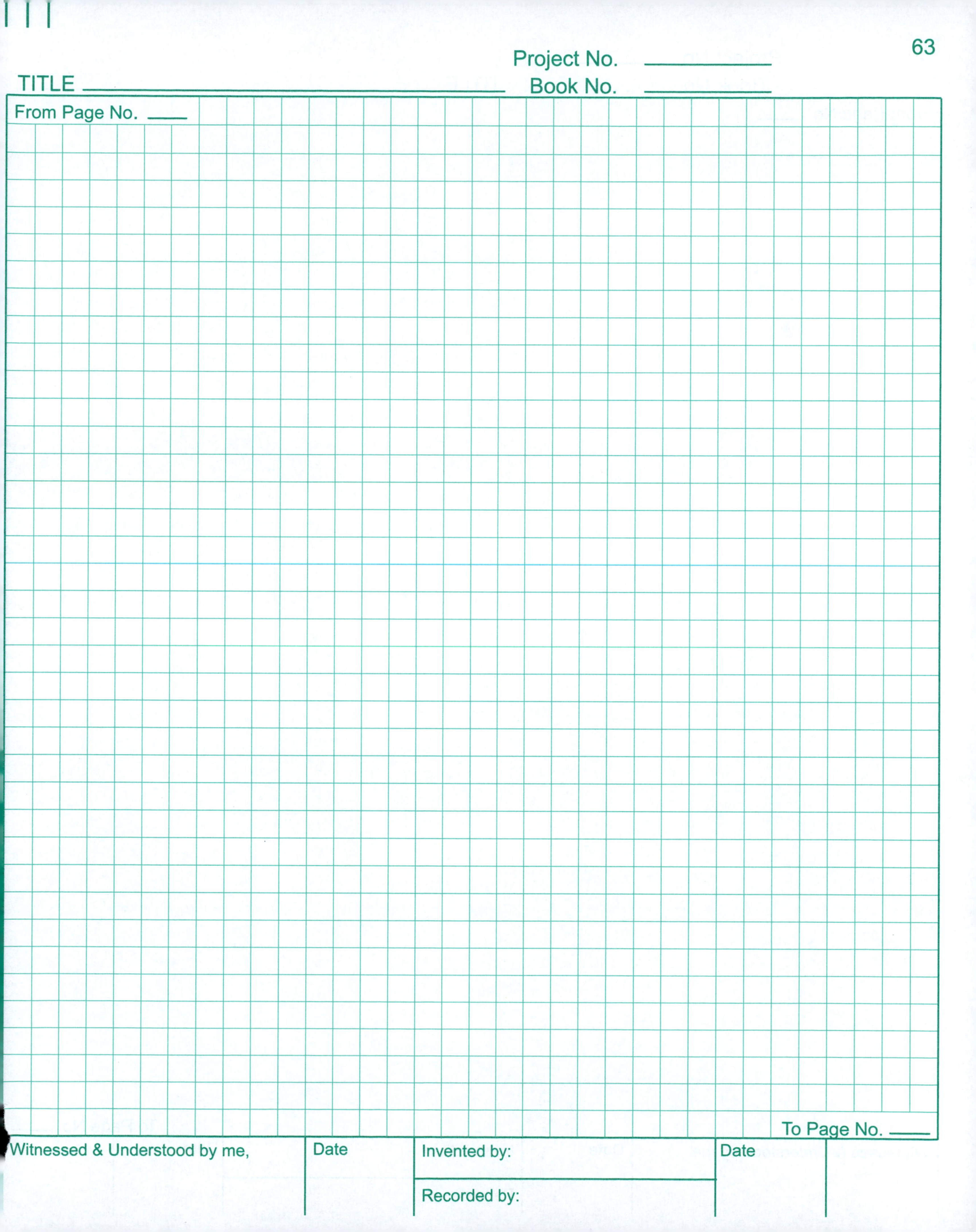

Project No. ____________

TITLE ____________ Book No. ____________

From Page No. ____

To Page No. ____

Witnessed & Understood by me,	Date	Invented by:	Date
		Recorded by:	

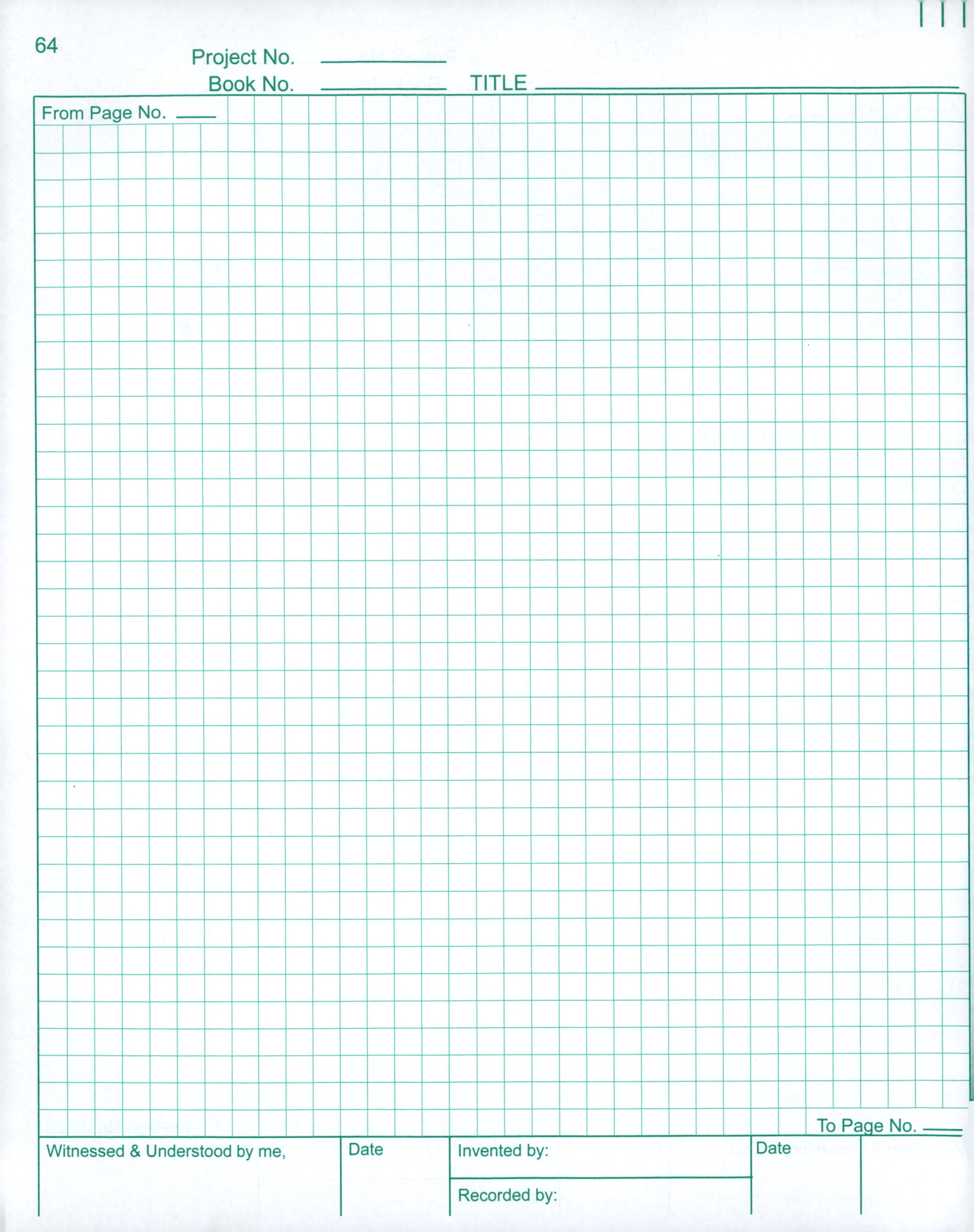
Project No.
Book No.
TITLE
From Page No.
To Page No.
Witnessed & Understood by me,
Date
Invented by:
Recorded by:
Date